International Nuclear Science Technology and Engineering Conference 2023

IOP Conference Series: Materials Science and Engineering
Volume 1308

Selangor, Malaysia
10-12 October 2023

ISBN: 978-1-7138-9717-0
ISSN: 1757-8981

Printed from e-media with permission by:

Curran Associates, Inc.
57 Morehouse Lane
Red Hook, NY 12571

Some format issues inherent in the e-media version may also appear in this print version.

Institute of Physics
Dirac House, Temple Back
Bristol BS1 6BE UK

Phone: 44 1 17 929 7481
Fax: 44 1 17 920 0979

techtracking@iop.org

Additional copies of this publication are available from:

Curran Associates, Inc.
57 Morehouse Lane
Red Hook, NY 12571 USA
Phone: 845-758-0400
Fax: 845-758-2633
Email: curran@proceedings.com
Web: www.proceedings.com

TABLE OF CONTENTS

IOP Conf. Series: Materials Science and Engineering 1308 (2024) 011001 doi:10.1088/1757-899X/1308/1/011001

IOP Conference Series: Materials Science and Engineering

International Nuclear Science, Technology and Engineering Conference 2023 (iNuSTEC2023)

"Nuclear Science and Technology For Socio-economic Development "

Universiti Tenaga Nasional
Kajang, Selangor, Malaysia
10-12 October 2023

Editors

Mohd Zamri Yusoff, Universiti Tenaga Nasional
Izham Zainal Abidin, Universiti Tenaga Nasional
Faridah Mohamad Idris, Malaysian Nuclear Agency
Julie Andrianny Murshidi, Malaysian Nuclear Agency
Hassan Mohamed, Universiti Tenaga Nasional
Mohd Syukri Yahya, Universiti Tenaga Nasional
Nuraslinda Anuar, Universiti Tenaga Nasional
Madihah Mujaini, Universiti Tenaga Nasional
Rogemah Ramli, Universiti Tenaga Nasional
Anas Muhamad Pauzi, Universiti Tenaga Nasional
Nurainaa Kabilmiharbi, Universiti Tenaga Nasional
Asyraf Arif Abu Bakar, Malaysian Nuclear Agency
Abdul Aziz Mohamed, Malaysian Nuclear Society

All papers have been peer-reviewed

Supporting/Sponsoring Organisations

Ministry of Education Malaysia; Ministry of Science, Technology and Innovation Malaysia; Malaysian Nuclear Agency; ASME Malaysia Section; Malaysian Welding and Joining Society; Reactor Interest Group; Asia-Oceania Neutron Scattering Association, Japan Atomic Energy Society; Institute of Nuclear Energy, Universiti Tenaga Nasional

IOP Conf. Series: Materials Science and Engineering 1308 (2024) 011001 doi:10.1088/1757-899X/1308/1/011001

MESSAGE FROM:
VICE-CHANCELLOR, UNIVERSITI TENAGA NASIONAL (UNITEN)

UNITEN is honored to host the International Nuclear Science, Technology, and Engineering Conference 2023 (iNUSTEC2023) and the Nuclear Youth Competition (NYC). With the theme "Nuclear Science and Technology for Sustainable Development Goals", iNuSTEC2023 is the thirteenth in a series of conferences hosted by the Malaysian Nuclear Society (MNS) since 2010 through collaboration with research institutes and local universities.

Don't we feel like time has passed by so quickly? And with it, came along significant and some were unthinkable changes to our lives. According to Our World in Data statistics, in 2022, each of us consumed on average 3.5 times more energy compared to 40 years ago.

In the year 2050, Malaysia is expected to consume almost twice the amount of electricity in 2022 (320 TWh vs. 180 TWh). This means that on a daily basis in 2050, Malaysia is expected to consume close to 0.9 TWh of electricity per day.

About 50 million electric vehicles are expected to be on the road in 2050. Every day, about 0.6 TWh of electricity is needed to charge these cars. That is an additional two-thirds of the expected daily electricity consumption per person in 2050.

Malaysia has identified the pathways for Responsible Transition 2050 in the National Energy Transition Roadmap (NETR) to tackle the energy trilemma of affordability, security, and sustainability. These pathways include having 0% coal share and 70% renewable energy share of installed capacity in 2050.

70% of renewable energy share is about 57 GW of solar PV, which will take a total land size that is slightly smaller than Melaka state. The same amount of power can be generated by nuclear power on a total land size that is 10 times smaller. That is not even considering the intermittency of power generation with solar PV.

At UNITEN, we believe that research and knowledge in nuclear power can play a huge role in supporting the Low Carbon Nation Aspiration and the Responsible Transition, especially in providing affordable, clean and safe energy supply to the Rakyat. Institute of Nuclear Energy (INE) at UNITEN was established in May this year to further empower and accelerate the existing research and educational efforts done at UNITEN since 2010.

At the university level, UNITEN has established the National Energy Centre (NEC) to support the national energy transition efforts. NEC will serve as a hub for research institutes, industrial and international linkages, and government agencies to form collaborations with experts from local and foreign universities, energy agencies, and stakeholders.

UNITEN envisions that NEC will play a significant role in providing technological solutions for Malaysia's energy transition which will be beneficial to Malaysia, as well as the surrounding region. NEC will also play the role of capacity building for the country's future workforce that will lead Malaysia towards net-zero in 2050. I extend my warmest invitation to every participant attending this conference, to be a part of this concerted nationwide effort in championing sustainability through involvement with the NEC.

I would like to remind participants, that you are among the very small percentage of Malaysia's population, who are experts in the nuclear field. Thank you so much for your contributions to keeping the knowledge related to nuclear science and engineering alive, despite how hard or negative it can get in the industry.

Congratulations to the organizing committee for hosting this iNuSTEC2023 together with the Nuclear Youth Competition (NYC2023). My deepest appreciation goes to our keynote speakers,

IOP Conf. Series: Materials Science and Engineering 1308 (2024) 011001 doi:10.1088/1757-899X/1308/1/011001

technical co-sponsors, and all supporters for their contribution to ensuring the success of this conference. To the presenters and participants, thank you for your participation and I wish the conference all success.

PROFESSOR IR. DR. NOOR AZUAN ABU OSMAN
Vice-Chancellor
Universiti Tenaga Nasional

IOP Conf. Series: Materials Science and Engineering 1308 (2024) 011001 doi:10.1088/1757-899X/1308/1/011001

MESSAGE FROM:
PRESIDENT, MALAYSIAN NUCLEAR SOCIETY (MNS)

I am really pleased to congratulate all keynote speakers, sponsors, presenters, and participants for the contribution and participating in the International Nuclear Science, Technology and Engineering Conference (iNuSTEC2023). Ideally, the iNuSTEC2023 is organized to enhance the importance of nuclear science, technology, and engineering in developing socio-economic in Malaysia. Malaysian Nuclear Society (MNS) is truly honored to organize this event for the third time with Universiti Tenaga Nasional (UNITEN). Successful conferences from previous years, at both national and international levels, have brought this event to be more meaningful to strengthen nuclear development for global growth under challenging situation

With the support from the American Society of Mechanical Engineers (ASME) Malaysia Section, and Malaysian Welding and Joining Society (MWJS), the MNS is indeed honored to co-organise this event with UNITEN and with organizations such as Nuclear Malaysia Agency, UTM, UKM, UM, UiTM, USIM and UTHM. The affiliation of the Institute of Nuclear Energy of UNITEN and advanced Nuclear Engineering Research Group of UTM and Institute of Nuclear Engineers (INE) of MNS, to make iNuSTEC2023 a success is a good testament that nuclear still remain relevant in the heart of the young generation to move towards greater strength in nuclear fields. In the future, MNS members shall strive to reach out to more diverse population in order to make nuclear ecosystem more conducive and stronger.

This conference had been supported by experienced and young researchers that can give a platform for them to discuss and debate on the latest development in all areas of nuclear technology applications for friendly purposes at the international level. This would spur and catalyse further research and development leading to the sustainability and improvement of the quality of life without compromising the quality of the environment in Malaysia and the world.

MNS is a unique non-governmental organisation that is formed and headed by individuals that are aligned to the beneficial and peaceful usage of nuclear science, engineering, and technology which has been incorporated under the Malaysian Society Act. The MNS has been actively promoting the information, advancement, popularization, and understanding of nuclear the public in Malaysia and this conference is one of the ways of realizing such aim. This conference is also served as a public acceptance program to introduce the benefits and advantages of the use of nuclear science, engineering, and technology to the public on a large scale. On top of that MNS also published twice a year on regular basis the Journal of Nuclear and Related Technologies (JNRT) and Buletin Nuklear Malaysia (BNM) for researchers to share their research findings with the community. We also publish iNuSTEC proceedings on yearly basis.

I would like to express my thank to UNITEN for hosting iNuSTEC2023 and Nuclear Youth Competition 2023. I wish to express my heartful thanks to all speakers for your willingness to share your experience with the participants and presenters for the development of the nuclear field. The contribution is not only coming from the Malaysian scientific community but also distinguished guest speakers of iNuSTEC2023. to all the personnel involved for their unwavering efforts to make this event a success. May all these efforts bring benefits to all of us and I congratulate everyone involved in making this conference successful event.

Associate Professor Dr Abdul Aziz Mohamed
President
Malaysian Nuclear Society

IOP Conf. Series: Materials Science and Engineering 1308 (2024) 011001 doi:10.1088/1757-899X/1308/1/011001

CONTENTS

PAPERS

IOP Conf. Series: Materials Science and Engineering 1308 (2024) 011001 doi:10.1088/1757-899X/1308/1/011001

iNuSTEC2023 ORGANISING COMMITTEE

***Patron*:**
Professor Ir. Dr. Noor Azuan Abu Osman
Vice Chancellor, Universiti Tenaga Nasional

Advisors:
- Prof. Dato' Ts Dr. Mohd Zamri Yusoff
  Deputy Vice Canselor (Research & Innovation), Universiti Tenaga Nasional
- Dr. Abdul Aziz Mohamed
  President, Malaysian Nuclear Society

Chairperson:
- Dr. Mohd Syukri Yahya
  Director, Institute of Nuclear Energy, Universiti Tenaga Nasional
- Prof. (Adjunct) Dr. Faridah Mohamad Idris
  Secretary, Malaysian Nuclear Society

Secretary:
- Ts. Dr. Nuraslinda Anuar, *Institute of Nuclear Energy, Universiti Tenaga Nasional*
- Ts. Asyraf Arif Abu Bakar, *Malaysian Nuclear Society/Malaysian Nuclear Agency*

Treasurer:
- Dr. Julie Andrianny Murshidi, *Malaysian Nuclear Society*
- Ms. Nurainaa Kamilharbi, *Institute of Nuclear Energy, Universiti Tenaga Nasional*

***NYC Chairs*:**
- Ms. Madihah Mujaini, *Institute of Nuclear Energy, Universiti Tenaga Nasional*
- Ms. Puteri Nuraliah Husna Mohd Tajuddin, *Malaysian Nuclear Society/Malaysian Nuclear Agency*

***Event Manager*:**
- Ms. Rogemah Ramli, *Institute of Nuclear Energy, Universiti Tenaga Nasional*

***Special Function Taskforce*:**
- Dr. Eli Syafiqah Azaman, *Institute of Nuclear Energy, Universiti Tenaga Nasional*
- Dr. Muhammad Arif Sazali, *Universiti Teknologi Malaysia*
- Ms. Khair'ah Binti Yazid@Khalid, *Malaysian Nuclear Society/Malaysian Nuclear Agency*
- Ts. Dr. Hassan Mohamed, *Institute of Nuclear Energy, Universiti Tenaga Nasional*
- Dr Muhammad Rawi Mohamed Zin, *Malaysian Nuclear Society/Malaysian Nuclear Agency*
- Mr. Azraf Azman, *Malaysian Nuclear Society/Malaysian Nuclear Agency*
- Ts. Razali Haron, *Malaysian Nuclear Society*

IOP Conf. Series: Materials Science and Engineering 1308 (2024) 011001 doi:10.1088/1757-899X/1308/1/011001

NUCLEAR YOUTH COMPETITION 2023 SECRETARIAT

Project Leader: Ms. Madihah Mujaini, *Institute of Nuclear Energy, Universiti Tenaga Nasional*

Deputy Project Leader: Ms. Puteri Nuraliah Husna Mohd Tajuddin,
Malaysian Nuclear Society/ Malaysian Nuclear Agency

Treasurer: Ts. Asyraf Arif Abu Bakar, *Malaysian Nuclear Society/ Malaysian Nuclear Agency*

Committee Members

Mr. Mohammad Amirudin Mohamad Rosli, *Malaysian Nuclear Society/ Malaysian Nuclear Agency*
Mr. Muhammad Zulhelmi Mahadi, *Malaysian Nuclear Society/ Malaysian Nuclear Agency*
Mr. Khairul Anwar Bapujee, *Malaysian Nuclear Society/ Malaysian Nuclear Agency*
Mr. Adam Haziq Mohamad Fahmi, *Universiti Teknologi Malaysia*
Ms. Ko Yu Yun, *Universiti Teknologi Malaysia*
Mr. Choo Kai Jie, *Universiti Teknologi Malaysia*
Mr. Syamil Ayman Suhailan, *Universiti Teknologi Malaysia*
Ms. Tang Jia Xuan, *Universiti Teknologi Malaysia*

Judges
Ts. Asyraf Arif Abu Bakar, *Malaysian Nuclear Society/ Malaysian Nuclear Agency*
Mr. Muhammad Ashraf Musauddin, *The eCEOs Sdn. Bhd.*
Mr. Khairul Izzat Abd Aziz, *Reltech Lab Sdn. Bhd.*
Ms. Zurina Ismail, *Yayasan Canselor Universiti Tenaga Nasional*

IOP Conf. Series: Materials Science and Engineering 1308 (2024) 011001 doi:10.1088/1757-899X/1308/1/011001

NUCLEAR YOUTH COMPETITION 2023

Champion:
Muhammad Mustaqim Bin Lokman
Nor Afiqah Binti Nor Asri
Nor Mazleen Binti Mokhtar
Nurul Auni Afifah Binti Ahmad Zahidi
Ho E-Sian

2nd Place:
Lee Kah Suan
Muhammad Amin Bin Mohd Nasir
Nor Irwanie Emeera Binti Mohd Nor
Nur Ain Izzati Binti Rosli

3rd Place:
Farahiyah Alyaa Binti Badrul Hisham
Seenevasan Naidu Pallie Naidu
Anis Salsabila Binti Ahmad Hafizullah
Mohd Azri Bin Ab Aziz

IOP Conf. Series: Materials Science and Engineering 1308 (2024) 011001 doi:10.1088/1757-899X/1308/1/011001

ACKNOWLEDGEMENT

iNuSTEC2023 is jointly organized by Malaysian Nuclear Society (MNS)
and Universiti Tenaga Nasional (UNITEN)

In Collaboration with

Malaysian Nuclear Agency (Nuklear Malaysia)
Universiti Teknologi Malaysia (UTM)
Universiti Kebangsaan Malaysia (UKM)
Universiti Teknologi MARA (UiTM)
Universiti Islam Antarabangsa Malaysia (IIUM)
Universiti Malaya (UM)
Universiti Sains Islam Malaysia (USIM)
Universiti Tun Hussien Onn Malaysia (UTHM)

Co-organized by
Institute of Nuclear Energy UNITEN
Advanced Nuclear Engineering Research Group UTM
Institute of Nuclear Engineers (INE MNS)
Malaysian Nuclear Professional Chapter (MNPC MNS)

Supported by
Reltech Lab Sdn. Bhd.
The eCEOs Sdn. Bhd.
Jimah East Power Sdn. Bhd
Yayasan Canselor UNITEN (YCU)

IOP Conf. Series: Materials Science and Engineering 1308 (2024) 011001 doi:10.1088/1757-899X/1308/1/011001

With Compliments

LIST OF EXHIBITORS/ SPONSORS

Exhibitors:

E1. **RELTECH LAB SDN BHD**

E2. **UNIVERSITI TENAGA NASIONAL**

E3. **MALAYSIAN NUCLEAR AGENCY**

E4. **MALAYSIAN NUCLEAR SOCIETY**

Sponsors:

- **TENAGA NASIONAL BERHAD**
- **YAYASAN CANSELOR UNITEN**
- **THE eCEOs SDN BHD**
- **RELTECH VENTURES SDN BHD**

IOP Conf. Series: Materials Science and Engineering 1308 (2024) 011001 doi:10.1088/1757-899X/1308/1/011001

IOP Conf. Series: Materials Science and Engineering 1308 (2024) 011001 doi:10.1088/1757-899X/1308/1/011001

MNS COUNCIL MEMBERS 2023/2025

Advisor / Past President	Prof. Dr. Nahrul Khair Bin Alang Md Rashid (*UTM*)
President	Dr. Abdul Aziz Bin Mohamed (*MNS*)
Vice President	Azraf Bin Azman (*Nuklear Malaysia*)
Secretary	Prof (Adj) Dr. Faridah Bt Mohamad Idris (*UNITEN/MNS*)
Deputy Secretary	Dr Mazleha Maskin (*Nuklear Malaysia*)
Treasurer	Dr. Julie Andrianny Bt. Murshidi (*Nuklear Malaysia*)
Deputy Treasurer	Dr Nazaratul Ashifa Binti Abdullah Salim (*Nuklear Malaysia*)

Members
Dr Mohd Idzat Idris (*UKM*)
Ts Asyraf Arif Abu Bakar (*Nuklear Malaysia*)
Dr Khairul Nadzmi Bin Jamaluddin (UTM)
Anwar Abdul Rahman (*Nuklear Malaysia*)
Dr. Mohd Syukri Yahya (*UNITEN*)

Auditors
Muhammad Amir Asyraf bin Ramli (*Arey Solutions*)
Zaredah Hashim (*Nuklear Malaysia*)

MNS Chapters
RadTech (*Radiation Tech*)
XApp (*X-ray App*)
NuEnergy (*Nuclear Energy*)
Neutron Photon (*Neutron- Photon*)
Student Chapter (*UKM, UiTM, UNITEN, UTM - NESS*)
Women in Nuclear, Malaysia (*WiN Malaysia*)
Institute of Nuclear Engineers (*INE MNS*)
Malaysian Nuclear Professional Chapter (*MNPC MNS*)

MNPC-MNS COMMITTEE MEMBERS 2022/2023
Advisor / President MNS: Dr. Abdul Aziz Bin Mohamed
President: Ts. Razali Bin Harun (*REM*)
Vice President: Khairul Izzat Bin Abdul Aziz (*Reltech Lab*)
Secretary: Mohd Taha Bin Hj. Suja' (*REM*)

INE MNS COMMITTEE MEMBERS 2022/2023
President: Asyraf Arif Bin Abu Bakar (Nuklear Malaysia)
Vice President: Khairul Anwar Bin Bapujee (Nuklear Malaysia)
Secretary: Puteri Nuraliah Husna Binti Mohd Tajuddin (Nuklear Malaysia)
Committee: Dr Muhammad Syahir Sarkawi (UTM),
Muhammad Zulhelmi Bin Mahadi (Nuklear Malaysia)

Visit us at http://www.nuklearmalaysia.org

Peer Review Statement

All papers published in this volume have been reviewed through processes administered by the Editors. Reviews were conducted by expert referees to the professional and scientific standards expected of a proceedings journal published by IOP Publishing.

1.Type of peer review: Single anonymous

2.Conference submission management system: Morressier

3.Number of submissions received: 24

4.Number of submissions sent for review: 24

5.Number of submissions accepted: 21

6.Acceptance Rate (Submissions Accepted / Submissions Received × 100): 91.7 %

7.Average number of reviews per paper: 2.54

8.Total number of reviewers involved: 29

9.Contact person for queries:

Name: Faridah Mohamad Idris

Affiliation: UNITEN

Email: faridah246@gmail.com

* means value has been edited

IOP Conf. Series: Materials Science and Engineering 1308 (2024) 012001 doi:10.1088/1757-899X/1308/1/012001

Single-layer Compton camera based on a 1 mm thick pixelated CDTE detectors

Lombigit L[1,a)], Maneuski D[2], and Bergmann B[3]

[1]Instrumentation & Automation Centre, Technical Support Division, Malaysian Nuclear Agency, Bangi 43000, Kajang, Selangor, Malaysia

[2]SUPA School of Physics and Astronomy, University of Glasgow, G12 8QQ, United Kingdom

[3]Institute of Experimental and Applied Physics, Czech Technical University in Prague, Husova 240/5, 110 00 Prague 1, Czech Republic

[a]lojius@nmgov.my

Abstract. Compton camera is a novel γ-camera paradigm that relies upon the kinematics of Compton scattering for image reconstruction. A Conventional Compton camera uses two layers of sensors - the Absorber and the Scatterer. This work reports the proof of concept of a single-layer Compton camera (SLCC) where simultaneous Compton pairs events are registered in a single High-Z semiconductor sensor. The Hybrid pixel detector (HPD) is 1 mm thick, 256 x 256 square pixels with 55 μm pixel pitch CdTe sensor bonded to a Timepix3 readout ASIC. The superior spectroscopic imaging and fast timing capabilities of the Timepix3 readout ASIC coupled with a microscopic and highly pixelated CdTe enable simultaneous event detection of multi-energies occurring at multiple positions. The concept was exemplified by measuring the 122 KeV γ-ray emitted from a 57Co radioisotope source at two positions. Data were captured in the Timepix3 data-driven mode with the KATHERINE readout system via Gigabyte Ethernet data transfer. A bespoke Compton kinematics criterion algorithm implemented in Python 3 IDE was used for data analysis and Compton's image reconstruction. Numerous events (5.2 million) were captured for 30-minute acquisitions. However, due to the thin nature of CdTe, fewer events (≈ 0.01 %) met the Compton event selection criteria. Nevertheless, the algorithm accurately pinpointed the radioisotope's location, demonstrating proof of concept of the SLCC system.

1. Introduction

Compton camera is a novel gamma camera concept that relies on Compton scattering kinematics for image reconstruction instead of physical collimation as in a conventional gamma camera system [1, 2, 3]. This feature gives the Compton cameras greater detection efficiency and a wider field of view. A conventional Compton camera system uses two detectors; the first detector acts as Scatterer, while the second is the Absorber. During the Compton scattering process, the recoil electrons deposit their energies in the scatterer detector, and the scattered photons that escaped the first detector are absorbed and deposit their energies through the photoelectric effect in the absorber detector. The single-layer Compton camera (SLCC) is a Compton camera variant where both the recoil electron and scattered photon were detected in a single detector system [4]. A hybrid pixel detector with a high-Z CdTe sensor bonded to a Timepix3 readout ASIC makes the SLCC possible. This is because the CdTe is sufficient to stop the Compton event in a single volume, and the Timepix3 permits simultaneous detection of energy and time occurrence of the event with high precision (Time of Arrival (ToA) = 1.562 ns).

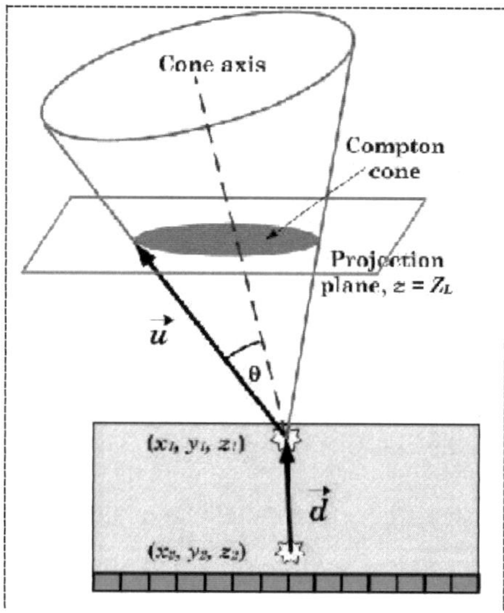

Figure 1. Image projection in a single-layer Compton camera (SLCC) system.

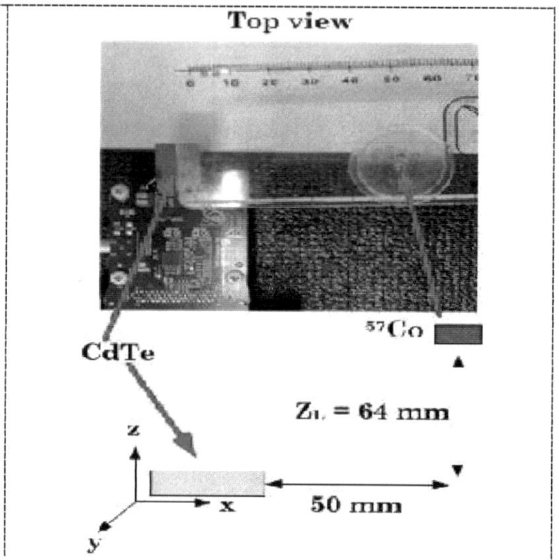

Figure 2. The experiment setup for the SLCC demonstration.

In an SLCC system (figure 1), Compton pairs deposit their energies in two 3D coordinates, i.e., scattering (x_1, y_1, z_1) and absorption (x_2, y_2, z_2) coordinates inside the detector volume. The imaging principle of an SLCC (see figure 1) relies upon the Compton scattering angle $\left(\theta = \arccos\left[1 - m_ec^2\left(\frac{1}{E_2} - \frac{1}{E_1+E_2}\right)\right]\right)$, where m_ec^2 = 511 keV is the electron rest mass energy whilst E_1, E_2 are the deposition energies of scattering and absorption events, respectively. Solving the scalar product of vectors connecting the absorption to the scattering coordinates for each Compton pair generates a Compton cone on the projection plane. Iterating the process for the identified valid Compton pair events produced many Compton cones, and the intersections of these cones represent the origin of the incident photon. Since the summed energy of scattering and absorption events is equal to the incident photon, thus, the radioisotope that produces the incident photons is also known. This paper presents the demonstration of an SLCC with a 1 mm thick CdTe detector coupled to a Timepix3 readout ASIC. The data acquisition system, the experiment setup, and the data processing method are explained, and the results will be discussed.

2. Materials and methods

2.1. Data acquisition system and experiment setup

Table 1. Data acquisition system and photon source used in the SLCC demonstration.

Hybrid pixel detector	Readout electronics & DAQ software	Photon source
1 mm thick CdTe + Timepix3 readout ASIC	Katherine Readout	1 Mbq ^{57}Co
No. of pixels: 256 x 256, 55 µm-pitch	Burdaman software	E = 122 keV
Bias voltage = -300 V	Interface: Gigabyte Ethernet	

Table 1 shows the data acquisition system and photon source used in this paper. The hybrid pixel detector is a 1 mm thick, 256 × 256 pixels with 55 µm-pitch CdTe bonded to a Timepix3 readout ASIC. The hybrid pixel detector is calibrated per pixel for energy and corrected for time anomalies (time-walk in the Time-of-Arrival (ToA) and column delay). Data comprising photon energy (ToT: Time-over-Threshold) and time occurrence of the event (ToA: Time-of-Arrival) were acquired in Timepix3's data-driven mode with the Katherine readout and Burdaman software [5]. Two experiments were conducted; *EXP 1*, the ^{57}Co was placed 50 mm in the x-axis away from the detector at a vertical distance, Z_L = 64 mm (figure 2) and *EXP 2*, where the ^{57}Co is placed 50mm below the detector with the same vertical distance. The acquisition time for the *EXP 1* and *EXP 2* are 3600 s and 1800 s, respectively.

2.2. Compton event selection criteria

Data were processed with an algorithm written in *Python 3* IDE utilising *Numpy, Pandas, Scipy* and *Matplotlib* packages for data handling, processing, and visualisation. Only events comprising two interactions that coincide within the clustering time were selected as potential candidates for the valid Compton events. The clustering time is the travel time of the electron-hole pair drifting across the detector thickness. For **this** case, the clustering time is assumed at around 32 ns, calculated from the electron mobility μ_e, bias voltage V_{bias} and detector thickness, d relationship $\left(\text{clustering time} = d^2/\mu_e V_{bias}\right)$. From these coincidence groups, energy cuts were implemented to select events with summed energy equal to incident photon energy:

$$\left(E_1 + E_2 = E_\gamma \pm {}^{\Delta E_\gamma}/_{E_\gamma}\right) \tag{1}$$

and exclude events belonging to the CdTe X-rays fluorescence lines:

$$\left(E \neq E_{XRF} \pm {}^{\Delta E_{XRF}}/_{E_{XRF}}\right). \tag{2}$$

The Compton edge:

$$\left(E_{Edge} = {}^{E_\gamma}\Big/\left(1 + \frac{m_e c^2}{E_\gamma}\right)\right) \tag{3}$$

was used to identify events belonging to the scattering (E_1) or absorption (E_2). The hybrid pixel detector provides information on the energy, time (ToA) and 2D pixel coordinate (x, y) of the Compton pair events. The scattering event was assumed to occur at the surface ($z_1 = 0$), and the z_2 is calculated from the depth difference between the two events ($z_2 = -1 *$ depth difference). The depth difference is given by $d\left(\frac{ToA_1 - ToA_2}{\tau_{Coll}}\right)$, where ToA_1, ToA_2 are the arrival time of E_1 and E_2. The τ_{Coll} is the time taken by the charge cloud to drift across the detector thickness, which is equal to the clustering time of 32 ns. Using the Compton scattering equation, the scattering angle θ was computed from the 3D coordinate of the Compton events. In addition to the energy cuts and XRF removal described earlier, two crucial criteria were implemented to obtain accurate image projection:

i. The scattering angle θ shall be within $0 < \theta$ [rad] < 1.0.
ii. A sizeable distance between the Compton event: 2D pixel distance ≥ 15 pixels.

Criteria (i) assumed the scattering event occurred at the top following the convention used in the traditional Compton camera system and (ii) to filter the XRF events.

2.3. Compton cone projection

In figure 1, vectors:

$$\vec{u} = (x - x_1,\ y - y_1,\ z - z_1) \tag{4}$$

and

$$\vec{d} = (x_1 - x_2,\ y_1 - y_2,\ z_1 - z_2) \tag{5}$$

connect the scattering and absorption events to the Compton cone. The 3D coordinate ($x, y, z = Z_L$) is the variable coordinate on the Compton cone surface, and the Compton cone is the scalar product of $\vec{u}$ and $\vec{d}$ given as:

$$\vec{u} \cdot \vec{d} = |\vec{u}||\vec{d}| \cos\theta. \tag{6}$$

Squaring Equation (6) yields:

$$(\vec{u} \cdot \vec{d})^2 = (\vec{u} \cdot \vec{u})(\vec{d} \cdot \vec{d})cos^2\theta = 0. \tag{7}$$

Equation (7) can be expressed into a second-order quadratic equation as:

$$(\vec{u} \cdot \vec{d})^2 - (\vec{u} \cdot \vec{u})(\vec{d} \cdot \vec{d})cos^2\theta = Ax^2 + Bxy + Cy^2 + Dy + Ey + F = 0, \tag{8}$$

where A, B, C, D, E and F are quadratic coefficients computed from 3D coordinates and scattering angle θ for each identified valid Compton events. The solution of equation (8) is a conic section representing the intersection between the Compton cone and the projection plane. Iterating this process for the identified Compton event produced many conic sections, and the intersection of these conic sections produced an image revealing the origin of the incident photon.

3. Result and discussion

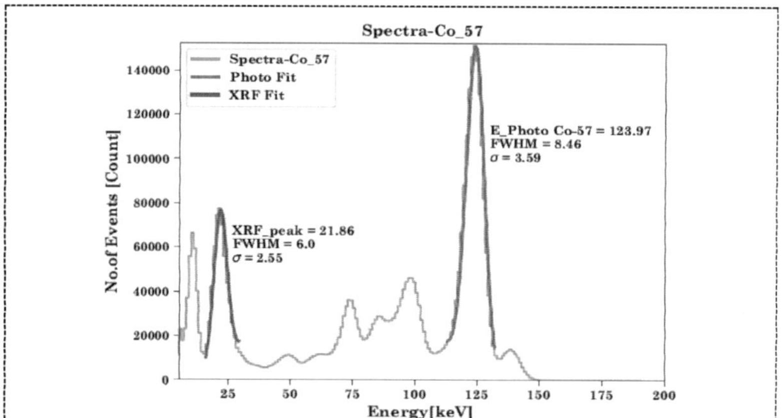

Figure 3. Energy spectra of ^{57}Co showing the main photopeak at 123.97 keV and the CdTe X-ray fluorescence around 21.86 keV.

Table 2. Analysis of number of events obtained from the two experiments.

	No. of events	
	EXP 1	*EXP 2*
Total	517, 9128	357, 7144
Coincidence events	8283	7823
Valid event	938	872
Final data	590	574

Figure 3 shows the energy spectra of ^{57}Co taken with the 1 mm CdTe bonded to Timepix3 readout ASIC. The photopeak of the ^{57}Co γ-rays is at 123.97 ± 8.46 keV, whereas the XRF peak is at 21.86 ± 6.00 keV. The energy resolution (FWHM) at respective peaks was used as the range of the energy and XRF cuts to select the valid Compton event. Table 2 shows the analysis of the number of events detected in both experiments. The 1 mm thick CdTe recorded many events, but mostly from the photoelectric interactions since this process is the dominant photon interaction at 122 keV in a Z = 48 CdTe sensor. Less than 0.2% (*EXP 1* = 8283 & *EXP 2* = 7823) were considered as candidates for valid Compton events as these events are two events that coincide within the 32 ns clustering time. Only ≈ 11% of the possible candidates pass the energy and XRF cuts, and ≈ 600 events passed the θ and 2D pixel distance criteria.

Figure 4 shows the reconstructed Compton images from the two experiments. Despite limited events, the source origin was sufficiently pinpointed in reconstructed images in both experiments. The reconstructed image's lack of sharpness was due to fewer data and the uncertainty in the θ measurement. Longer measurement time and (or) thicker sensors would improve the data statistics. The detector' s resolution (spatial and energy) and the Doppler broadening affect the θ measurement. The spatial resolution generates divergence between the measured θ and actual θ, which can be higher than 6° for a 55 μm-pitch and at 50 mm source to detector distance [6]. The energy resolution and Doppler broadening affect the accuracy of the energy measurement. In this paper, the detector was biased directly from the Katherine readout, which is maximum at −300 V. The detector is specified for −500 V; therefore, the obtained resolution is not optimal. The Doppler broadening occurs in the energy spectra of the scattering and absorption events. This effect is due to the incident photon interacting with moving electrons that are bound to atoms [7].

IOP Conf. Series: Materials Science and Engineering 1308 (2024) 012001 doi:10.1088/1757-899X/1308/1/012001

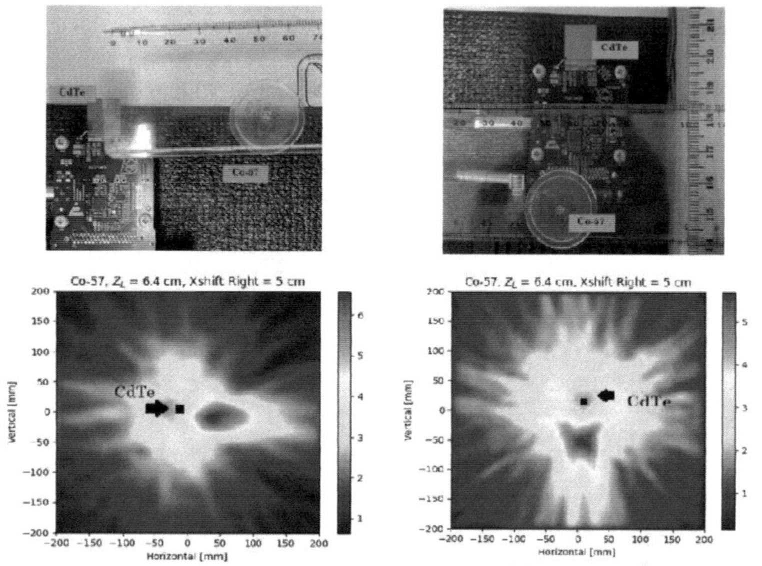

(a) Source at 5 cm - right (b) Source at 5 cm - bottom

Figure 4. The reconstructed Compton image from the two experiments.

4. Conclusion

A demonstration of a single-layer Compton camera system with a 1mm thick CdTe bonded to a Timepix3 readout ASIC irradiated with 122 keV γ-rays from 57Co has been presented. Due to the thinness of the CdTe sensor used, only a limited number of valid Compton events were obtained as the photoelectric absorption is the dominant interaction at 122 keV. A careful selection of the scattering angle and a large separation between the two coincidence events were used besides the energy and XRF cuts to improve the accuracy of the image projection.

5. References

[1] Turecek D, Jakubek J, Trojanova E, and Sefc L 2018 *J. Instrum.* **12** C1102.
[2] Gutierrez A *et al.* 2018 *J. Instrum.* **13** C01036.
[3] Sakai M *et al.* 2019 *Sci. Rep.* **9** 12906.
[4] Turecek D, Jakubek J, Trojanova E and Sefc L 2020 *J. Instrum.* **15** C01014.
[5] Burian P *et al.* 2017 *J. Instrum.* **12** C11001.
[6] Wu C *et al.* 2022 *Phys. Med. Biol.* **67** 125017
[7] Ordonez C E, Bolozdynya A and Chang W 1997 *IEEE Nucl Sci Symp Conf Rec.* **2** 1361-1365.

Acknowledgement

The study is sponsored by Malaysia's Public Service Department (JPA) under the Federal Training Grant (HLP) and travel assistance allowance from the University of Glasgow Lord Kelvin Mobility Fund. Also, acknowledge the staff at the Institute of Experimental and Applied Physics (IEAP), CTU of Prague, for their assistance.

IOP Conf. Series: Materials Science and Engineering 1308 (2024) 012002 doi:10.1088/1757-899X/1308/1/012002

Optimization of oil recovery using heated low salinity water (HLSW) in the horizontal sand pack column during water flooding: radiotracer intervention

Noraishah Othman[*,1], Akram Mujaddid Bin Abdul Halim[2], Muhammad Firdaus Hakimi bin Mohd Zulkarnain[3], Nazrul Hizam Yusoff[1]

[1]Industrial Technology Division, Malaysian Nuclear Agency, 43000, Kajang, Selangor, Malaysia
[2]Faculty of Science and Technology, Universiti Sains Islam Malaysia (USIM), Bandar Baru Nilai, 71800, Nilai, Negeri Sembilan, Malaysia
[3]Fakulti Sains Gunaan, Universiti Teknologi MARA Perlis, 02600, Arau, Perlis, Malaysia

*noraishah@nm.gov.my

Abstract. Water-flooding is a prevalent technique for secondary oil recovery that is capable to increase oil recovery by up to 45%. Due to its accessibility, affordability, and simplicity, water flooding is the most frequently used secondary oil recovery technique. However, the efficiency of conventional water flooding is limited by capillary forces which holds the residual crude oil in pore structures. Moreover, during water flooding operation, the injected fluid does not penetrate sufficiently due to low permeability or presence of channelling inside the sandstone reservoir. Therefore, in this study, thermal recovery was introduced to 100ppm (0.1g/L) low salinity water (LSW) in order to investigate the percentage of oil recovery during water flooding. A horizontal sand pack column was used for water flooding experiment and temperature and injections rate are the parameters to be optimised with respect to percentage of oil recovery. The column was initially packed with 125micrometer sieved sand to ensure constant homogeneity is achieved. Initially, formation water which comprised with NaCl, CaCl2 and MgSO4 was introduced inside the column together with light oil and was aged overnight. TAPIS oil with viscosity of 0.001382 Pa.s (1.382cP) was used as light-oil in this study. The injection rate was set to 1 mL/min, 2 mL/min, and 3 mL/min respectively whereas, the temperature of the LSW was heated at 50 °C, 70 °C, and 90 °C subsequently. Heated water was supplied continuously throughout the sand pack which was heated using water blanket to retain heat at 70°C (reservoir temperature). In conclusion, 3 mL/min and 70°C were identified as optimum parameters and the oil recovery experiment was proceeded with liquid radiotracer using Technecium-99m (Tc-99m) intervention and resulted with 73% of yield. Nevertheless, the radiotracer intervention only provides the Residence Time Distribution (RTD) model which describe the behaviour of the sand pack during water flooding process.

1. Introduction

Water flooding is one of the most utilized Enhanced Oil Recovery program in order to extract oil in place from matured reservoir. Recently, low high salinity water (LSW) has attracted attention of the stake holders and numerous researches have been carried out to optimize the yield of oil from the oil reservoir. Previous authors claimed that LSW is able to extract additional of 5-38% from the reservoir unlike high salinity water that cannot tolerate with the clay particles attached to the pore walls due to wettability issues thus hinder the oil recovery [1]. It is believed that the kaolinite from the pore walls, which is the clay particles will detach easily by the presence of LSW. The oil that is clinging together with moveable kaolinite clay at the pore walls will be easily swept thus, enhances the recovery of oil. According to [2], the recovery of oil is significantly enhanced when the salinity of injecting water for water flooding activity is lower than connate water. Moreover, LSW injection is most preferred because of its cost effectiveness due to the availability of water and cheaper than other approach plus the oil displacement is slightly easier due to its excellent maneuverability in oil reservoir. Seawater, fresh water and injecting water from other reservoirs or produced water are the most common water sources for water flooding operations [3]. However, water flooding is best applied to light crude oil reservoir. The crude oil that we used for this study is from TAPIS reservoir which has viscosity of 0.001382 Pa.s (1.382cP) and considered as light oil. Light crude oil is defined as having low density and viscosity and move freely in room temperature.

Malaysia reservoir is basically fractured basement rock reservoir located offshore. The structures are characterized by a fracture system of very high heterogeneity that leads to anisotropic flow [5]. Thus, radiotracer technology is proposed due to its ability to permeate through any voids and soil stratification. This is due to the fact that liquid tracer can squeeze through any soil openings that leads to information of the reservoir as well as displacing the residual oil. Basically, analysis of residence time distribution (RTD) is carried out in order to determine model that fits the reservoir well before thorough interpretation of the respective reservoir can be made. Hence, the primary objective of this study is to optimize parameter in enhancing oil recovery for sandstone reservoir. Thus, heated low salinity water (HLSW) is proposed to substitute conventional LSW water flooding in order to investigate the effectiveness of thermal expansion during water-flooding activity. In this study, the sand column is heated at 70°C throughout the experiment to mimic the temperature of actual reservoir. Moreover, this study aims to investigate the hydrodynamics behavior inside the porous domain by providing the analysis of residence time distribution (RTD) models using radiotracer intervention.

2. Methodology

2.1. Water flooding set-up: optimization parameters

Figure 1 shows the schematic diagram of overall experiment for optimization parameters. The experiment was started with an empty column compacted with 125 micron -sized sand. The fully loaded column was weighed to obtain the mass of sand in the rig. Next, formation water was prepared as shown in Table 1 and injected inside the column using peristaltic pump followed by 100 ml of kerosene subsequently and left for a day. Brine solution was prepared inside a beaker and heated up to designated temperature. Both beaker and column were wrapped with aluminum foil to maintain the temperature. Finally, then brine solution was injected inside the column using peristaltic pump at designated flow rate for 6 hours. The injection rate was set to 1 mL/min, 2 mL/min, and 3 mL/min respectively whereas, the temperature of the LSW was heated at 50 °C, 70 °C, and 90 °C accordingly.

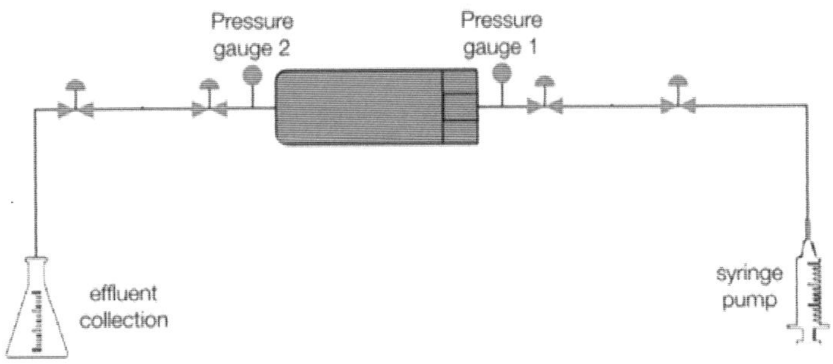

Figure 1. Schematic diagram of experimental set up.

Table 1. Composition of formation water and low salinity water.

Chemicals	Formation Water	Brine Solution (100ppm)
NaCl(g)	28.295	0.1
CaCl(g)	0.887	-
$MgSO_4$	0.079	-
Distilled Water (L)	1	1
Tapis Oil (viscosity)	0.001382 Pa.s (1.382cP)	

2.2. *Radiotracer experiments*

Radiotracer experiment was carried out on the elevated core flood rig after optimum parameters (temperature and flow rate) were obtained from 2.1. The radioactive experiment was conducted once in order to reduce the radiation exposure to the operator as well as for safety purpose. Similar procedure was adopted as 2.1 except in this experiment 6 sodium iodide (Tl) scintillation detectors were arranged externally along the column. All detectors were connected to Data Acquisition System (DAS) which was connected to laptop for LSW-tracer monitoring. Technetium-99m, a gamma emitter water tracer with 0.140MeV was used for water flooding activity. About 5.37mCi/3ml of Tc-99m was injected inside the column while LSW was running continuously.

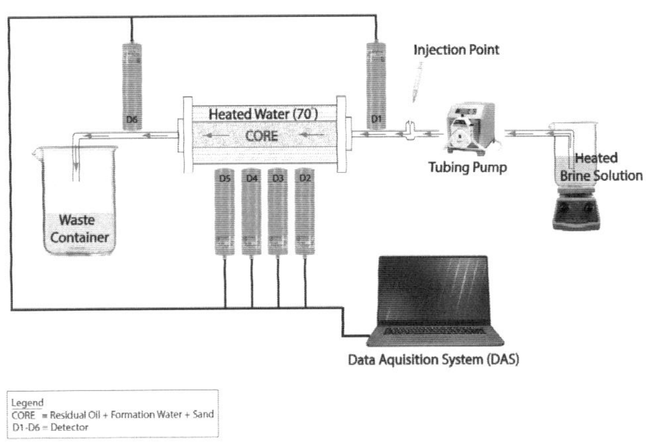

Figure 2. Schematic diagram of radiotracer experiment.

IOP Conf. Series: Materials Science and Engineering 1308 (2024) 012002 doi:10.1088/1757-899X/1308/1/012002

3. Results and discussions

3.1. Optimization of water flooding parameters

In this study, the injection rate was set to 1 mL/min, 2 mL/min, and 3 mL/min respectively whereas, the temperature of the LSW was varied at 50 °C, 70 °C, and 90 °C accordingly. According to Afiq *et al.* [3], when injection rate increases, the capillary force is increased that leads to high sweep efficiency and improved oil displacement. The existence of fingering is also avoided by selecting the lower flow rate for water flooding injection. Besides, the low viscosity light oil is used which reduces the presence of viscous fingering. Similarly, as temperature increases, the yield is significantly improved. This is due to the fact that hot water injection is a thermal recovery technique which reduces significantly the viscosity of oil. Thus, by thinning the oil viscosity it provides the driving mechanism for the hot water to push the residual oil freely.

Moreover, the Low Salinity Water provides additional mechanism to detach the oil from the sand wall easily hence improve oil displacement and achieve greater oil recovery result. This is because the interaction of ions from the water phase with carboxylic acid from the oil phase leads to changes in interfacial tension (IFT) of the fluids. From the compiled findings [4], whereby the salinity under investigations were 15000-100000ppm, it was concluded that IFT increases with increasing salinity. Thus, since 100ppm is used for the study, the IFT between LSW-oil is decreased and the recovery of oil is significantly increased as compared to previous authors as shown in Table 2. In conclusion, the oil recovery is proportional to temperature and flow rate in the range of temperature and flow rate used in this study as shown in Figure 3. However, the optimized parameters were set at 3ml and 70°C since the actual reservoir temperature is 70°C.

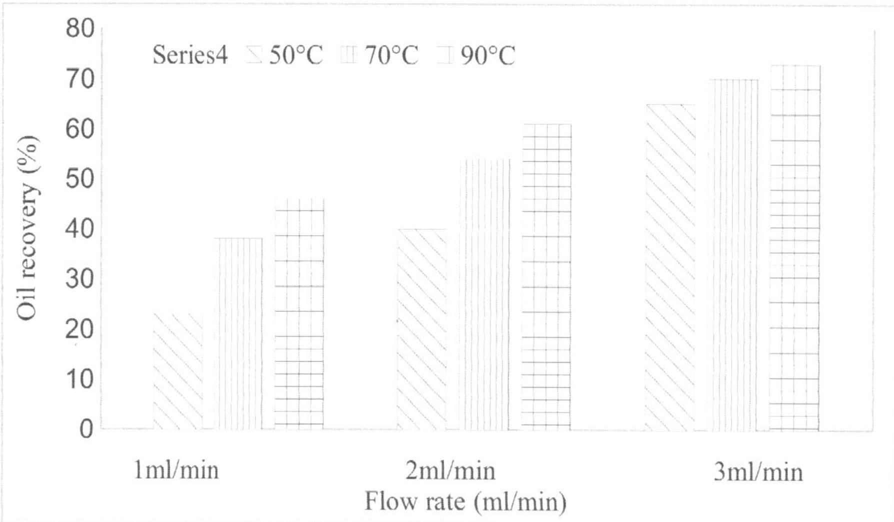

Figure 3. Parameters of process optimization.

IOP Conf. Series: Materials Science and Engineering 1308 (2024) 012002 doi:10.1088/1757-899X/1308/1/012002

Table 2. Oil recovery from various salinity concentration.

Authors	Remarks	Arrangement of column	Original Oil In Place (OOIP) (%)
Present study	Heated Low Salinity Water (HLSW) (100ppm)	Horizontal	73.0
Noraishah Othman et al. (2022)	300 ppm	Vertical	62.1
N.Othman et al. (2021)	500 ppm	Vertical	66.7
Danial Azim Che Aziz et al. (2020)	15 wt% kaolinite concentration	Horizontal	55%
Noraishah Othman (2020)	30,000 ppm brine solution	Horizontal	38.37%

3.2. Radiotracer experiments: qualitative analysis

Figure 4 shows the radioactive signals acquired by each detector once the water flooding commences. The continuous water injection carries Tc-99m which emits gamma rays and the scintillation detector from individual detector is picking up the signal. D1 portrays highest and sharp peak which implies the radioactive material has highest concentration and sufficient strength prior entering the column. Similarly, the peak is still significant at D2 because the mixing of HLSW-tracer is about to start inside the sand column and the pressure is relatively low. As the HLSW-tracer moves from right to the left of column, the resistance inside the column is very dominant. The pressure is increasing since the HLSW-tracer has to loose itself from the packed sand, formation water and residual oil respectively in order to permeate and distribute homogenously inside the packed column and finally towards the end of the column. Therefore, the distribution of the curve from D2 until D5 is gradually broaden and flattened. Moreover, it can be inferred that D1 is producing high level of counts compared to all detection sensors because there is no resistance experience by D1 since the detection is carried out outside of the sand column. Figure 1 indicates that all detectors are able to pick up radioactive signals which are due to fluid flow from water-flooding activity of HLSW-tracer inside the column.

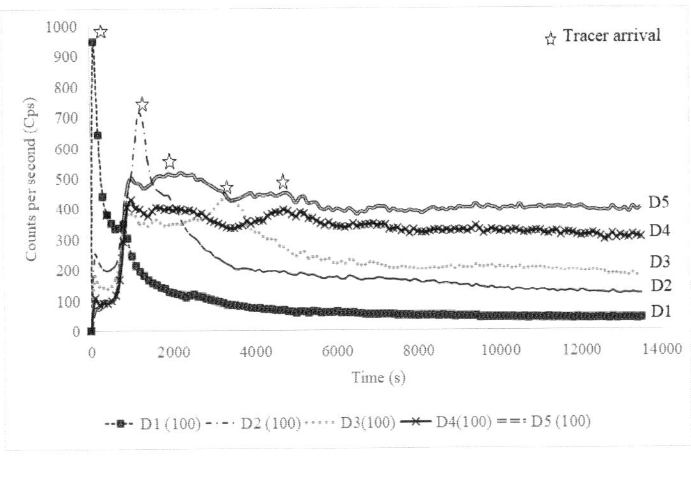

Figure 4. Radiotracer experiments: Fluid flow of HLSW-tracer inside the sand pack column.

3.3. Radiotracer experiments: quantitative analysis

In order to determine the process anomalies inside the sand pack during water flooding, analysis of residence time distribution (RTD) is carried out. Basically, assigned RTD model enables to provide the deviant of the respective process from ideal condition. The International Atomic Energy Agency (IAEA) has recommended six RTD models to be fitted with experimental RTD derived from radiotracer experiment [9]. Each model has its own parameters that is optimized in order to select the best fitting curve. Hence, the sum of square errors (SSE) is used as means to justify the optimum RTD model to tracer response experiment. The least value of SSE, the minimum deviation of between the experimental data and the model function, is the benchmark for model selection among the recommended RTDs [10].The mathematical expression for RTD is shown in Equation (1). The E(t) function is the RTD where E(t) dt is the fraction of the flow, measured at the exit, that is in the system between times t and (t +dt) [9]:

$$E(t) = \frac{c(t)}{\int_0^\infty c(t)\,dt} \tag{1}$$

where C(t) is the concentration of radiotracer monitored by NaI scintillation detectors in counts per second (cps) as numerator and denominator is the area under the curve of plotted C(t). Figure 5 shows the data from Detector 6 which is the response of the tracer concentration profile at the outlet of the sand pack and the concentration measurements are recorded using data acquisition system (DAS) as shown in Figure 2. This data is rich in information that provides the information of process behavior during the water flooding process. The data is then transformed using Eq. [1] and the RTD models against experimental data are plotted using IAEA RTD Software as shown in Figure 6. From observation, the perfect mixer in series with exchange is the most representative model than other models because the SSE is 0.364 x 10^{-9}, the least value compared to other models as shown in Table 1.

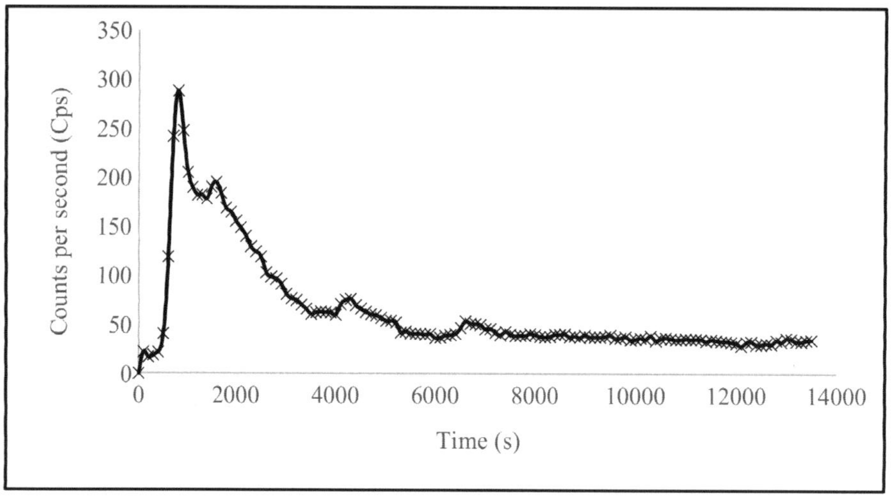

Figure 5: Raw data acquired from D-6 (outlet of the column)

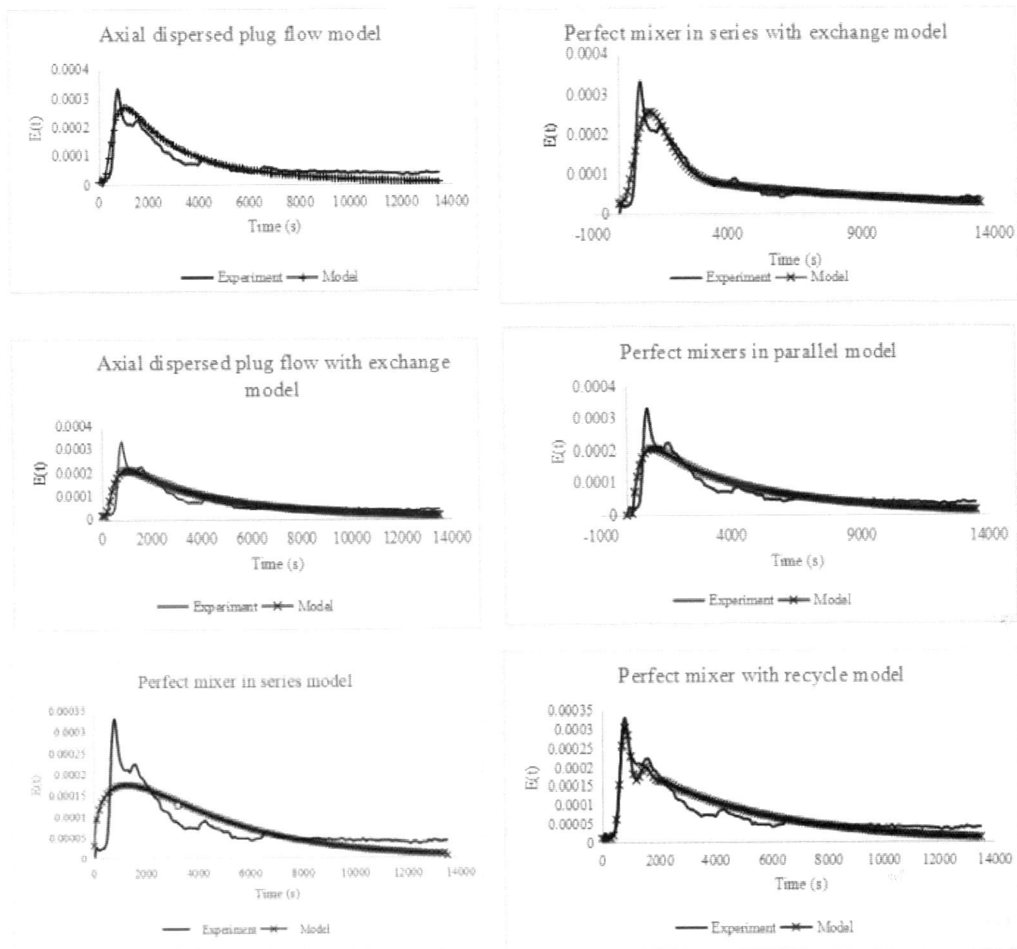

Figure 6. RTD Models with respect to IAEA recommendation

Table 3. Sum of Squares Sum of Squares Error

Models	Parameters									SSE
	Pe	N	τ_1	τ_2	J_1	J_2	T_m	K	Q_1/Q_2	
Axial dispersed plug flow	1.5		4636							0.762×10^{-9}
Axial dispersed plug flow with exchange	1.1	1.0	2910		1.0					0.753×10^{-9}
Perfect mixers in series			4325		1.4					1.46×10^{-9}
Perfect mixers in series			1824		4.4		0.48×10^4	3.3		0.364×10^{-9}

IOP Conf. Series: Materials Science and Engineering 1308 (2024) 012002 doi:10.1088/1757-899X/1308/1/012002

with exchange							
Perfect mixers in parallel	4434	313	0.73	1.0		1.2	0.805 x10^{-9}
Perfect mixers with recycle	8571	-173	185	- 0.32 x10^{7}		5.9	0.463 x10^{-9}

*τ=mean residence time, J= Number of tanks, Pe=Peclet number, k=V_1/V_2, T_m=residence time for exchange zone

The perfect mixers in series with exchange consists of several parameters which are τ, J, T_m and K where all the variables are defined as follows:

$$\tau=\frac{V_1}{Q}\, T_m=\frac{V_2}{\alpha QJ}\, K=\frac{V_1}{V_2} \tag{2}$$

In this model, the mechanism of flow can be described in Figure 7. The flow is basically distributed into two distinctive volumes; V_1 which is total volume of the primary stream and V_2 total volume of secondary stream respectively. The secondary flow comprises with volume of exchange cells with interconnected αQ flow rate. As shown in Table 3, the number of perfect mixers of the respective column is 4 with mean residence time of 1824s for primary zone and 4800s for secondary zone respectively. The high value of T_m can be translated to the migration of ions from primary to secondary flow. K value is 3.3 which indicates the volume of primary zone is superior three times than secondary zone and thus, reflects the occurrence of an exchange flow rate αQ between the main flow and secondary flow whereas Q is the volumetric flow rate of the primary stream from one volume to another. The presence of liquid ions in formation water (Na^+, Cl^-, Ca^{2+}, Mg^{2+} , SO_4^-), brine (NaCl) and tracer (TcO_4^{-}) contribute to the exchange of ions between primary stream and secondary stream. Moreover, the interaction of HLSW- tracer has reduced the IFT and perhaps altered the wettability of sand towards water-wet, thus increase the oil production. Wettability alteration is required to ensure the wetting state of rock is dominant towards water wet since wettability alteration is an effective approach to enhanced oil recovery.

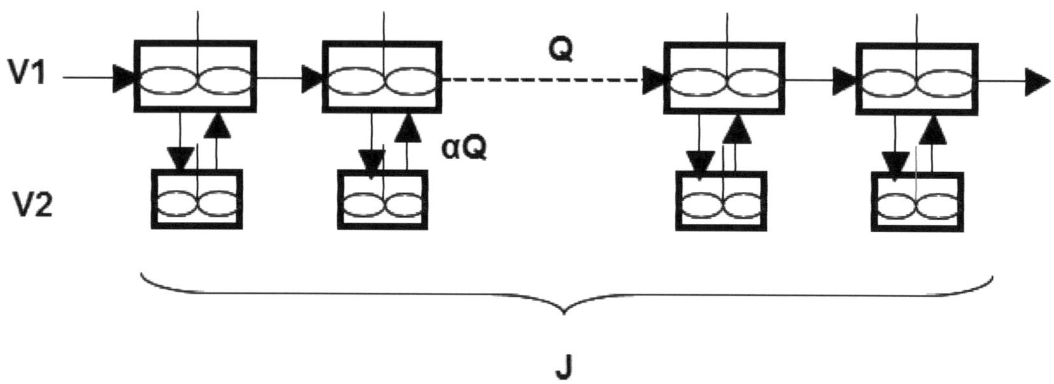

Figure 7. Flow mechanism of perfect mixers with exchange model [9].

4. Conclusions

In this study, two main analysis were carried out successfully which were optimization of residual oil from the heated low salinity water (HLSW) and determination of RTD model that represents the response of the tracer concentration profile at the outlet of the sand pack. Basically, the chosen RTD model describes the behavior or characteristics of the respective column. In this study, the optimum parameters for oil recovery is 3 mL/min and 70°C with yield of 73% of oil. This oil recovery is the highest in comparison to oil recovery from previous authors. Moreover, the perfect mixers with exchange model describes the sand column well due to the observed value of SSE is the minimum among all recommended models. The model defines then presence of four perfect mixers and has two distinctive volumes which are primary and secondary volumes. Both volumes are connected with an exchange flow rate denoted as αQ and the presence of ions are basically from formation water, brine and tracer respectively.

5. References

[1] Wan Zaidel W M A B, Rosli N R, Mohammadian E, Othman N, Yusoff N H 2022 Simulating the effect of salinity and kaolinite concentration on low salinity water flooding in sandstone reservoir of Sabah basin, *Mater. Today: Proc.* **63** S467-S476, https://doi.org/10.1016/j.matpr.2022.04.130

[2] Che Aziz D A, Mohammadian E, Rosli N R, Yusoff N H, Othman N 2020 The Effect of kaolinite on oil recovery by low salinity waterflooding in sandstone reservoirs *Int. J. of Mech. and Prod. Eng. Res. and Dev. (IJMPERD)* **10**(3) 15533–15540 ISSN (P): 2249–6890; ISSN (E): 2249–8001 (TJPRC Pvt. Ltd).

[3] Abdul Aziz A A, Md Yusop N, Othman N 2022 Influence of injection rate and oil viscosity on viscous fingering –simulation prediction *CFD Lett.* **14**(6) 24-423e.

[4] Sauerer B, Al-Hamad M, Ma S M, Abdallah W 2021 Effect of formation water salinity on interfacial tension of reservoir fluids *J. of Petrol. Sci.and Eng.* **204**.

[5] Noraishah Othman, Hearie Hassan, Nur Syawal Aiman, RTD Modelling Using Radiotracer Technique in Porous Media for Oil Recovery Study. AIP Conference Proceedings 2295, 020010 (2020); https://doi.org/10.1063/5.0031516

[6] Othman N, Jasni N S, Rosli N R, Yusof N H, Shari M R, Hassan H, Mahmood A A, Terry A M, Abdullah N 2022 Residence time distribution (RTD) model of water flooding in vertical sand column using Technetium-99m as the radiotracer *Malaysian J. of Chem. Eng. & Technol.* **5**(2) 99–106.

[7] Othman N, Wan Kamal W H B, Jasni N S, Rosli N R, Yahya R, Yusoff N H, Mohd Yunos M A S, Shah Dahing L N, Shari M R, Hassan H, Mahmood A A, Terry A M, Abdullah N 2021 Residence time distribution (RTD) of water flooding activity in oil recovery using vertical sand column: radiotracer approach *IOP Conf. Series: Mater. Sci. and Eng.* **1106** 012031 IOP Publishing doi:10.1088/1757-899X/1106/1/012031.

[8] Othman N, Saaid I M, Ahmed A, Yusof N H, Yahya R, Mohd Yunos M A S, Engku Chik E M F, S R, Hassan H, Mahmood A A 2020 Investigation of water-flooding activity using radiotracer technology in commercial core-flood set up *AJChE* 20(1) 49 – 56, doi: 10.22146/ajche.51902.

[9] IAEA 2008 Residence time distribution method for industrial and environmental applications, *Training course series 31* (Vienna, Austria).

[10] Sugiharto S, Suud Z, Kurniadi R, Wibisono W, Abidin Z 2009 Radiotracer method for residence time distribution study in multi-phase flow system. *Appl. Radiat. Isotopes* **67** 1445–1448.

Acknowledgement
The authors would like to thank all PAT colleagues who have directly and indirectly participated in this work. Special appreciation to the sponsors: International Atomic Energy Agency (IAEA) Research Contract No: 22898, IAEA Coordinated Research Project F22069 and Project No. FRGS/1/2018/TK07/MOSTI/02/3 provided by the Ministry of Higher Education Malaysia.

IOP Conf. Series: Materials Science and Engineering 1308 (2024) 012003 doi:10.1088/1757-899X/1308/1/012003

Probabilistic approach of solving burnup problems

K A Bala[1,a)], **M R Omar**[1, a)], **K F Chin**[1]

[1]School of Physics, Universiti Sains Malaysia, 11800 USM, Penang, Malaysia.

[a]rabieomar@usm.my

Abstract. This paper presents a new approach to solve nuclear burnup problems using a probabilistic method. Unlike the traditional methods that rely on complex matrix exponential calculations, the proposed method tracks the transformation of nuclei over a period of time, enabling the estimation of nuclide concentrations. The method is implemented in a C++ program *CNUCTRAN* and verified against the Chebyshev Rational Approximation Method (CRAM). Three sample calculations are presented, showing a detailed comparison of results obtained using *CNUCTRAN* and CRAM48. The relative error analysis emphasizes the accuracy of the probabilistic method, with the most significant error observed below 0.001%. The computational efficiency is discussed in terms of CPU time, indicating that *CNUCTRAN* requires more CPU time than CRAM but with improved accuracy as the time step decreases. The numerical results for various burnup problems demonstrate the accuracy and efficiency of the probabilistic method, making it a promising alternative for simulating realistic nuclear transmutations.

1. Introduction

Analyzing nuclear reactors requires careful consideration of how radioactive materials change over time and in response to nuclear reactions, for instance, decay, neutron capture and fission. The burnup equation describes the behaviour of the material composition over time in a nuclear reactor, which depends on the reactor's core parameters and material compositions [1]. The equation is usually a set of first-order linear differential equations. These equations are stiff, especially when the nuclides included in the calculation combine short-lived and long-lived ones. The Chebyshev Rational Approximation Method (CRAM) [2] and Transmutation Trajectory Analysis (TTA) [3] are two methods that earlier academics used to address challenging situations [4], and both are computationally efficient but are mathematically abstract. On the other hand, [5] recently found an unconventional way of solving burnup problems by deriving a probability distribution that predicts whether a nucleus will transform into another species. This paper presents a new non-deterministic approach to solve burnup problems, and the method is implemented in a C++ program named *CNUCTRAN* to verify the method.

2. Theoretical Background

2.1. Burnup equation

The burnup equation can be expressed in Eq. (1) below.

$$\frac{d_{y_i}}{dt}=\sum_{j=1}^{I} b_{ij}\lambda_j y_j+\sum_{k=1}^{I} f_{ik} A_{ik} y_k-\left(\lambda_i+A_i\right) y_i \qquad (1)$$

where y_i is the density of nuclide *i*, λ_i is the decay constant of nuclide *i*, b_{ij} is the branching ratio from nuclide *j* into nuclide *i*, f_{ik} is the production yield from nuclide *k* into nuclide *i* and $A_{ik}$is the production rate of nuclide *i* due to the removal of nuclide *k* from the system. Eq. (1) is a system of *I* linear differential equations dependent on one another. The solution of Eq. (1) is the concentration of *I* nuclides. The term on the left-hand side of Eq. (1) is known as the rate of change of the nuclide-*i* concentration. The left-hand side of Eq. (1) is balanced with the terms on the right-hand side. Here, each term describes the various transmutations of nuclides that lead to the production and removal of nuclide-*i* within the system.

In particular, the first term on the right-hand side of Eq. (1) represents the natural decay of various nuclides that transform into nuclide-*i*. In contrast, the second term of the right-hand side of Eq. (1) describes the formation of nuclides-*i* via different transmutation reactions of all nuclides, excluding the natural decay reaction. Finally, the last term on the right-hand side of Eq. (1) is a combination of natural decay and mixed transmutation reactions that transform nuclide-*i* into other nuclides, which removes it from the system. Alternatively, Eq. (1) can be expressed in its compact matrix form given by,

$$\frac{dy}{dt}=Ay \qquad (2)$$

where A is now known as the transmutation matrix. Also, A is a sparse $I \times I$ Square matrix, where most elements are zero, and y is a column vector containing the instantaneous concentration of all *I* nuclides defined in the calculation. Technically, the norm of A is in order of 10^{21}, and includes both off-diagonal elements and diagonal elements. However, using the separation of variables Eq. (2) can also be expressed as given below:

$$y(t)=e^{At}y_0 \qquad (3)$$

Alas, evaluating the matrix exponential in Eq. (3) proves to be a complex task. One might attempt to simplify the process by employing a Taylor series exponential or Padè approximation of the matrix exponential, which is unfortunately impractical and does not always lead to accurate and reliable results, especially for stiff burnup problems.

Pusa [6] observed that the eigenvalues of matrix A are situated in close proximity to the negative real axis. This allows for the estimation of the matrix exponential using Chebyshev approximants. This approach is commonly referred to as the Chebyshev Rational Approximation Method (CRAM), and it provides an approximation for the matrix exponential as follows:

$$e^{At}=a_0+2\Re ¿ \qquad (4)$$

where a_i and θ_i are the complex numbers published by Pusa [6] for the approximation order $k\in\{16,32,48\}$.

3. The probabilistic method

In the previous study, the establishment of $\pi_{ij}(\Delta t)$ (probability distribution) allows us to estimate the probability of various transmutation events experienced by a nucleus. The primary objective of this work is to estimate the time evolution of nuclide concentrations without directly solving Eq. (1) mathematically. This endeavour is underpinned by the utilization of the probability distribution given by [7],

$$\pi_{ij}(\Delta t)=\gamma_i \prod_{l=1}^{J_i} (\delta_{jl}+(-1)^{\delta_{jl}} ¿ e^{-\Lambda_{ij\Delta t}}) ¿ \quad (5)$$

where γ_i is the normalization constant and δ_{jl} is the Kronecker delta. Also, Λ_{ij} is the reaction rate constant of reaction j of nuclide i, and it corresponds to λ and A in Eq. (1). The value of γ_i is given by,

$$\gamma_i = \left(\sum_{j=0}^{J_i} \pi_{ij} \right)^{-1} \quad (6)$$

Generally, the concentration of nuclides i at a given time $t+\Delta t$, is contributed by two factors. Firstly, it is contributed by the un-reacted nuclides i within the interval $(t, t+\Delta t)$. Secondly, it is contributed by the transmutation of other nuclides that occur within the interval $(t, t+\Delta t)$, which leads to the production of nuclide i. Therefore, the concentration of nuclide i after the next sub-step, $t+\Delta t$, can be expressed as,

$$y_i(t+\Delta t)=\pi_{i \to i} y_i(t)+\sum_{\forall\, l \neq i} \pi_{l \to i} y_l(t) \quad (7)$$

where $\pi_{i \to i}$ is the probability of nuclide i not undergoing a transmutation event; $\pi_{l \to i}$ is the probability of any nuclides l undergoing a transmutation event that transforms themselves into nuclide i . By definition,

$$\pi_{l \to i} = \sum_{\forall\, j \in R} \pi_j \quad (8)$$

where R is a set of transmutation events that transform nuclide l into daughter nuclide i. If removal event j is a fission reaction, then π_j in Eq. (8) must be scaled to the fission yield of the daughter nuclide i. Eq. (7) can be systematically written in matrix form via the definition given in Eq. (8). Here, by considering all I nuclides in a burnup problem,

$$\begin{pmatrix} y_1(t+\Delta t) \\ y_2(t+\Delta t) \\ \vdots \\ y_I(t+\Delta t) \end{pmatrix} = \begin{pmatrix} \pi_{1 \to 1} & \cdots & \pi_{I \to 1} \\ \pi_{1 \to 2} & \cdots & \pi_{I \to 2} \\ \vdots & \ddots & \vdots \\ \pi_{1 \to I} & \cdots & \pi_{I \to I} \end{pmatrix} \begin{pmatrix} y_1(t) \\ y_2(t) \\ \vdots \\ y_I(t) \end{pmatrix} \quad (9)$$

Or in a simple form,

$$y(t+\Delta t)=P(\Delta t)\, y(t) \quad (10)$$

Notice that all elements in P are calculated based on the sub-step interval, Δt. Eq. (10) is a recursion relation where the future nuclide concentrations, $y(t+\Delta t)$, can be obtained by transforming the nuclide concentrations of the previous time interval via the transfer matrix, P. Here, P serves as the transformation operator. The column indices of P correspond to the various parent nuclides, and the row indices of **P** correspond to the different daughter nuclides. Notice that the diagonal elements of square matrix P are probabilities of no transmutation events occurring. The P element can be calculated using Eq. (7), provided all reaction rates associated with the nuclides are known.

Suppose that the evolution of nuclides concentration is monitored from the initial time, $t=0$, to the final time, $t=T$. Then, a small-time interval, Δt, is chosen such that the total number of sub-steps is given by $l=T/\Delta t$. Therefore, $y(t)$ can be obtained by repeating Eq. (10), for l times, giving,

$$y(T)=y(l\Delta t)=P^l y(0) \tag{11}$$

where $y(0)\equiv y_0$ is the initial nuclides concentration at $t=0$. Here, the smaller the value of Δt, the larger the number of sub-steps, thus the better the accuracy [7]. However, such a computation with small Δt requires arbitrary precision arithmetic to avoid the floating-point error. The evaluation of sparse matrix exponentiation, P^l, can be accomplished at an incredible speed via exponentiation by squaring. Evaluating P^l requires l repeated multiplications of P. Such a naïve approach will incur an unreasonably long CPU time, especially when l is large. Gratefully, the exponentiation by squaring helps to evaluate P^l with only $\log_2 l$ Multiplications [8]. The trick is that P^l can be efficiently calculated by repeated squaring of **P.** Since evaluating P^l also involving repeated matrix multiplications, the conventional matrix multiplication algorithm will impose unnecessary operations on zero elements of the sparse matrix, P. Therefore, this work also implements the sparse matrix multiplication (SpMM) algorithm proposed by [9].

Remember that the probabilistic method requires l self-multiplications of the transfer matrix P. in the previous section, such a computation can be done at an incredible speed via exponentiation by squaring. To save computational time, the number of substeps is chosen as

$$l=2^k \tag{12}$$

where k is the total number of self-multiplications of P. Our objective is to obtain the suitable value of k, such that the substeps interval Δt is in order of 10^{-n}. Here, n is known as the approximation order, and the time step, T, of the calculation is given below:

$$T\approx 2^k \Delta t=2^k 10^{-n} \tag{13}$$

Solving for k,

$$k=\log_2 10^n T=\lfloor \log_2 10^n T \rfloor \tag{14}$$

The floor operation in Eq. (13) is needed because k must be a positive integer number. Also, Δt must be corrected to compensate for the error due to floor operation:

$$\Delta t=\frac{T}{2\lfloor \log_2 10^n ¿}¿ \tag{15}$$

4. Sample calculations

The proposed method is implemented in a C++ program named *CNUCTRAN* and has been verified against the state-of-the-art method, CRAM. The nuclides data, such as the half-lives and other transmutation chains of over 4000 nuclides, are fetched from an XML file containing the transmutation information of various nuclides derived from ENDFBVII.1. The results obtained were compared with CRAM of order 48 (CRAM48). Table 1 summarizes all the information of sample 1,2 and 3 calculations.

IOP Conf. Series: Materials Science and Engineering 1308 (2024) 012003 doi:10.1088/1757-899X/1308/1/012003

Table 1: Sample calculations.

Sample	Total Nuclides	Parent Nuclides Involved	Initial Concentrations (mol)	Reaction rates (sec^{-1})	Time Step (sec)
1	1489 (A=150-250)	U-235, U-238	U-235: 0.1, U-238: 0.9	(n, a): 10^{-4}, (n, γ): 10^{-4}, fission: 10^{-5})	8.64×10^4
2	1603 (A=150-264)	U-235, U-238	U-235: 0.1, U-238: 0.9	(n, a): 10^{-4}, (n, γ): 10^{-4}, fission: 10^{-5})	8.64×10^6
3	1632 (A=137-237)	Np-237	Np-237: 1.0	Pure decay	3.16×10^{10}

4.1. Sample Calculation I: decay of U238 and U235

The first sample calculation is a decay reaction with a total of 1489 nuclides that involve U-235 and U-238 with initial concentrations of 0.1 mol and 0.9 mol, respectively. The system of nuclides is under a constant burnup such that the (n, a), (n,γ) and fission reaction rates are kept constant at $10^{-4} sec^{-1}$, $10^{-4} sec^{-1}$ and $10^{-5} sec^{-1}$. The sample monitored the concentrations of all nuclides involved for the time step of 8.64E4 sec. The results obtained from the proposed method exhibit negligible differences with the reference solution. Interestingly, 1249 nuclides exhibited zero concentrations in both the proposed methods, while 224 nuclides maintained non-zero concentrations. Additionally, 16 displayed negative CRAM concentrations, coinciding with zero concentrations in the proposed method. Notably, the relative error between solutions computed using *CNUCTRAN* and CRAM48 is illustrated in Fig. 1, offering a visual representation of the comparative accuracy of these computational methods.

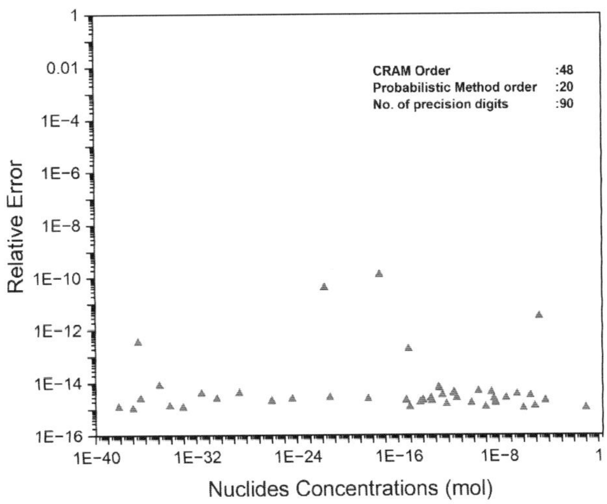

Figure 1. Relative error versus sample 1 after 8.64×10^4 sec.

4.2. Sample Calculation II: U238 and U235.

The second sample calculation is similar to the first sample which tracks 1603 total number of nuclides over the time step of 8.64×10^6 sec (100 days). The results obtained show 224 nuclides with non-zero concentration. 25 nuclides with negative CRAM concentration, and these nuclides have zero

IOP Conf. Series: Materials Science and Engineering 1308 (2024) 012003 doi:10.1088/1757-899X/1308/1/012003

concentration on the proposed method. 1354 nuclides with zero concentration on both two methods. A comprehensive assessment of the relative error between solutions computed using *CNUCTRAN* and CRAM48 is presented in Fig. 2. This graphical representation illustrates the comparative accuracy of these computational methods and provides a visual insight into their performance across the spectrum of tracked nuclides.

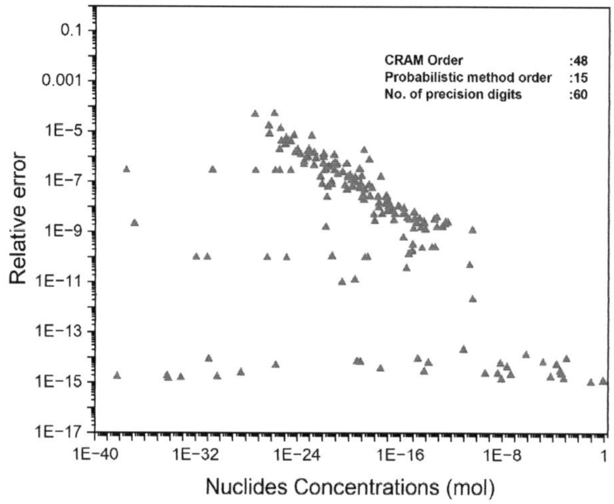

Figure 2. Relative error versus *sample II* after 8.64×10^6 sec.

4.3. Sample Calculation III: pure decay of Np-237

This sample calculation involves a pure decay reaction consisting of the linear decay chain of Np-237 with an initial concentration of 1 mol, and the nuclide concentrations were monitored over the time steps of 3.154×10^{10} sec. The sample involved 1632 total number of nuclides starting from atomic mass number of A =137 to A = 237. The obtained results shed light on the behaviour of these nuclides, revealing 15 with non-zero concentrations. Interestingly, most 1617 nuclides exhibit zero concentrations in the CRAM and *CNUCTRAN*, underlining a consistent behaviour across these computational approaches. To provide a quantitative perspective on the precision of the solutions, the relative error between computation using *CNUCTRAN* and CRAM48 is shown in Fig. 3. This graphical representation serves as a visual aid, offering a clear comparison of the accuracy of these computational methods in modelling the dynamic evolution of nuclides concentrations.

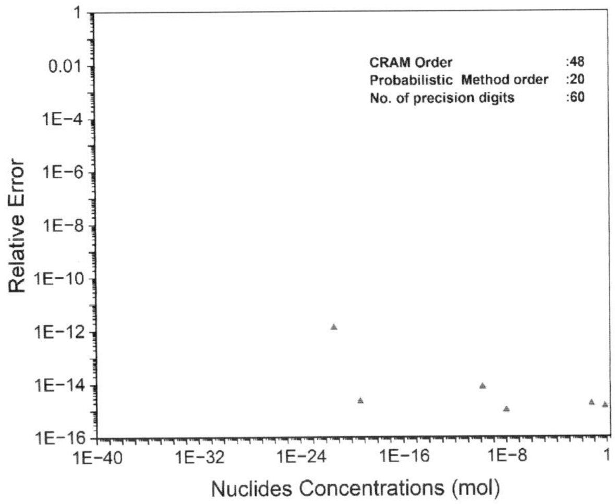

Figure 3: Relative error versus *sample III* after 3.154×10^{10} seconds.

Table 2. CPU Time (in seconds) comparison between *CNUCTRAN* and CRAM48.

Time step, t (s)	CPU Time (*CNUCTRAN* @2.20GHz, 2 cores)	CPU Time (CRAM, @2.20GHz, 2 cores)
8.64×10^4	1.274	0.781
8.64×10^6	1.294	0.828
3.154×10^{10}	1.200	0.672

The results obtained from the three distinct samples exhibit no significant differences from the reference solution, and the largest error observed is less than 0.001%. It is worth highlighting that the overall accuracy of *CNUCTRAN* improves as the time step size decreases. Therefore, this phenomenon is a known limitation of the probabilistic method outlined in section 3. It is important to note that only nuclides with concentrations 10^{-40} in three samples, respectively, were considered to minimize rounding errors. Additionally, the relative error increases with decreasing nuclide concentrations. Furthermore, 100% of the important nuclides have a relative error of less than 10^{-5}, indicating the outstanding accuracy of *CNUCTRAN*. For the benefit of readers, the CPU times for *CNUCTRAN* and the CRAM method are in Table 2. It is noteworthy that *CNUCTRAN* requires more CPU time than the CRAM method; as discussed in section 3, the CPU time of *CNUCTRAN* increases as the time step length increases.

5. Conclusion

In summary, the probabilistic method used an unconventional way to simulate the time evolution of nuclide concentrations. It stems from the direct simulation of nuclides transformations, and the implementation requires elementary probabilistic theories that can be easily understood. The presented method enables the inclusion of short-lived and long-lived nuclides in a single calculation. Even though it can stimulate realistic nuclide transmutations, it has a few limitations that necessitate future development. For instance, dealing with slowly evolving concentrations of long-lived nuclides requires arbitrary precision arithmetic to account for such a subtle change. Following that, the method requires a huge number of sparse matrix multiplications. Here, the speed of these multiplications depends on the computing. Further research and implementation of this method should focus on improving the speed of arbitrary precision sparse matrix multiplication.

6. References

[1] Bateman H 2023 The solution of a system of differential equations occurring in the theory of radioactive transformations Proc. Cambridge Philos. Soc. 15 423–427, 1910 Accessed: Nov. 01, 2023 Online https://cir.nii.ac.jp/crid/1571135649245258368.bib?lang=en

[2] Calvin O, Schunert S and Ganapol B 2021 Global error analysis of the Chebyshev rational approximation method *Ann. Nucl. Energy* **150**, doi: 10.1016/j.anucene.2020.107828.

[3] Cetnar J, Stanisz P and Oettingen M 2021 Linear chain method for numerical modelling of burnup systems *Energies (Basel)* **14**(6) doi: 10.3390/en14061520.

[4] Isotalo A E and Aarnio P A Comparison of depletion algorithms for large systems of nuclides *Ann. Nucl. Energy* **38**(2–3) b 261–268 Feb. 2011 doi: 10.1016/j.anucene.2010.10.019.

[5] Omar M R and Karim J A 2022 The probability distribution of transmutation events and its application for solving burnup problems *Progress in Nucl. Energy* **152** doi: 10.1016/j.pnucene.2022.104395.

[6] Pusa M 2016 Higher-order Chebyshev rational approximation method and application to burnup equations *Nucl. Sci. and Eng.* **182**(3) 297–318 doi: 10.13182/NSE15-26.

[7] Omar M R and Karim J A 2022 The probability distribution of transmutation events and its application for solving burnup problems *Progr. in Nucl. Ener.* **152** doi: 10.1016/j.pnucene.2022.104395.

[8] Gordon D M 1998 A Survey of fast exponentiation methods.

[9] Gustavson F G 1978 Two fast algorithms for sparse matrices: multiplication and permuted transposition *ACM Transact. on Math. Software (TOMS)* **4**(3) 250–269 doi: 10.1145/355791.355796.

Acknowledgements

This work was supported by the Ministry of Higher Education Malaysia through the Fundamental Research Grant Scheme [Project Code: FRGS/1/2022/STG07/USM/02/9].

Initial findings of autonomous radiation mapping robot field test

Muhammad Izzuan Mohd Ghazali[1,a)], Nur Aira Abd Rahman[1], Nabilah Ramli[1], Suzilawati Muhd Sarowi[1], Muhamad Zahidee Taat[1], Ahmad Hasnulhadi[1], Syirrazie Che Soh[1], Lojius Lombigit,1 Nor Arymaswati Abdullah[1], Noor Farhana Husna A Aziz[1], Mohd Shafiq Sazali[1], Azimawati Ahmad[1], Khairul Salleh[1]

[1]Malaysian Nuclear Agency, 43000, Bangi, Selangor, Malaysia

[a]izzuan@nm.gov.my

Abstract. Radiation mapping using mobile robotic can minimize the risk of operator exposure to radiation while providing detailed radiation maps of targeted areas. Despite its potential advantages, the use of this technology is still limited and requires further investigation. In a previous study, an autonomous radiation mapping robot was designed and developed. This paper presents the initial findings on autonomous radiation mapping field test that focuses on the operation procedures and the challenges faced during the robot deployment. Temporary solutions implemented to overcome the challenges are also discussed. The results will provide insights for future researchers and operators working on mobile robotic technology for radiation mapping.

1. Introduction

Radiation mapping simplifies the task of visually representing the distribution of radiation levels in an area. Specific devices designed to detect radiation levels are employed to measure the amount of radiation present in a particular location. These devices can vary and may include Geiger-Muller counters, scintillation detectors, or advanced monitoring instruments. They have the capability to identify and measure different types of radiation, including alpha particles, beta particles, gamma rays, and X-rays. The collected data is then analysed and transformed into a map that uses different colours or contour lines to indicate different radiation intensities. Instead of manual radiation measurement, robotic radiation mapping uses specialized robots with radiation detection sensors to create maps of radiation levels. These robots operate independently, gathering data in hazardous areas where human presence is risky. The collected data is then analysed to generate radiation distribution maps.

Robotic radiation mapping offers improved safety, efficiency, and accuracy compared to manual methods. Through the process of radiation mapping, it becomes feasible to evaluate the dangers associated with radiation, pinpoint potential origins of contamination, keep track of fluctuations in radiation levels over a period, and take necessary actions to reduce risks and safeguard the well-being of the public and the environment. The findings are reported to authorities for informed decision-making on safety measures. There were many endeavors have been attempted in order to reduce the risk of radiation exposure by using different method and robot in order to achieve their goals. For ground-based robotic platform, radiation detector normally mounted on a commercial or custom-made Unmanned Ground Vehicle (UGV) depending on which type of radiation wanted to be detected such as Jackal by Clearpath Robotics [1] suitable for uneven terrain and Thovald robot platform [3] which is used for agriculture purpose. As for determining the quality of air or finding traces and radiation reading of an open space, Unmanned Aerial Vehicle (UAV) such as quarter-scale Piper Cub J3 model airplane [2], Orpheus X-3 UGV carrying gamma radiation detectors, and DJI S800 UAS equipped with a multisensor system for aerial photogrammetry [5] can be used as according to which parameter and situation favorable to the process.

The objective of this paper is to present the procedure that we have developed to deploy mobile robot for radiation mapping experimental evaluations in selected MNA facilities. Subsequently, the challenges, issues, and solutions will also be described in this paper.

2. Methodology

2.1. Robot description

This project is called Autonomous Mobile Robot for Accurate Radiation Detection and Mapping which also known as AMoRA which is shown in Figure 1. In this study, the research utilizes the Turtlebot2 as the selected mobile robot platform. Kobuki is used as its base which also known for low cost and easy to use for education and research purposes. Kinect sensor is mounted on the robot which enable the robot to have the working range of 1.2 meters up to 3.5 meters range with 57° wide horizontal field of view or 3.8 meters wide at its maximum capacity and 43° wide or 63 cm of vertical field of view. This robotic system features a pair of differential drive wheels and is equipped with various navigation sensors, including an onboard RGBD camera, wheel encoders, gyro sensors, wheel drop sensors, bump sensors, and cliff sensors. The Turtlebot2 is paired with the Robot Operating System (ROS), an open-source framework licensed under the BSD license. Simulated robot navigation codes are executed within ROS Gazebo, and real-time data visualization is performed using ROS RViz.

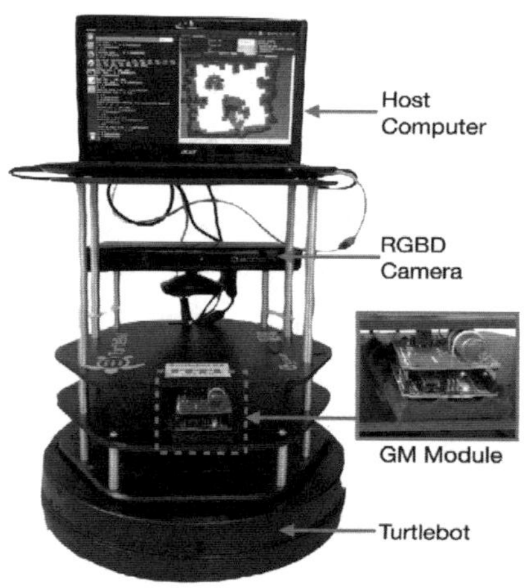

Figure 1: AMoRA.

The radiation detection system in this study utilizes the Geiger Muller (GM) detector LND7121, and our team previously developed the electronics module for this detector [15]. This module includes the GM detector itself, a 500-volt high-voltage supply for the detector, and a signal processing unit that converts the detector's analog output into a digital signal (TTL pulse). To tally the pulses generated by the GM detector, a counter timer is integrated into the microcontroller. The module's output provides a gross count of radiation events, measured in counts per second (CPS). As for connectivity, we link the GM electronics module to the AMoRA host computer through a USB connection, as depicted in Figure 2. This module is physically installed on one of the shelves of the AMoRA. To enable communication with the AMoRA 's ROS environment, we employ the ROS package *rosserial_arduino* to publish the radiation detector's output data on the */RadiationData* topic, with a publishing rate of 1 Hertz.

2.2 Radiation mapping procedures:

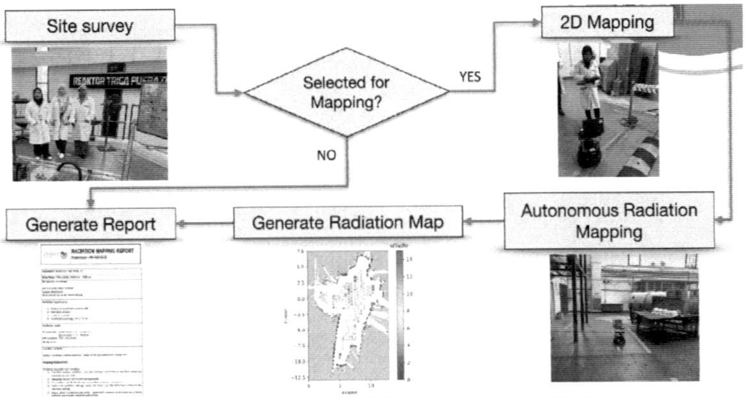

Figure 2: Flow Chart of Autonomous radiation mapping

2.2.1. Site survey. The initial phase of automated radiation mapping commences with site survey. This phase is significance due to its influence on the selection of appropriate sites compatible with the current robot capabilities. Several critical aspects warrant consideration during the survey:

a) *Terrain and surface condition:* The site should ideally feature a smooth and preferably indoor surface. This selection is necessitated by the characteristics of the AMoRA, which relies on dual wheels, thus imposing constraints on terrain. Additionally, a smooth surface is instrumental in maintaining the accuracy of the final manually generated map. Irregular or unstable surfaces can result in data inaccuracies and overlap issues.

b) *Obstacle assessment:* Careful assessment of obstacles is imperative, as the robot's sensors can only detect objects positioned 20 cm above ground level, owing to the placement of the RGBD camera at the same height. Consequently, objects lower than 20 cm remain unregistered, rendering the camera's line of sight obstruction-free below this threshold.

c) *Site structure:* A symmetrical arrangement or room structure can perplex the robot and the mapping program, leading to data overlap. This, in turn, necessitates multiple attempts and editing to ensure precise data acquisition as shown in Figure 3.

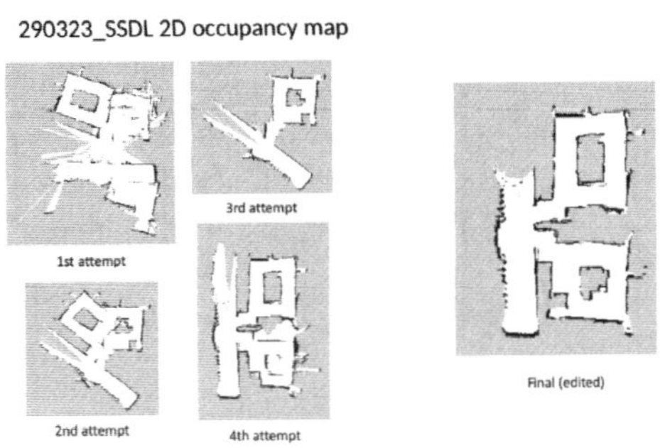

Figure 3. Multiple attempts of manual mapping.

2.2.2 2D mapping. Once the site selection is finalized, the process of 2D mapping or manual mapping can be initiated. This involves several sequential steps:

a) *Connecting robot and host computer:* Establishing a connection between the robot and the host computer, followed by launching a terminal to insert command prompts.
b) *Utilizing input devices:* Employing wireless or Bluetooth mouse and keyboard for long-distance control during manual mapping.
c) *Command prompts:* Executing a series of command prompts to establish connectivity, activate the Simultaneous Localization and Mapping (SLAM) algorithm, and run RViz, a visualization tool facilitating data interpretation from various sensors.
d) *Optimizing data clarity:* If map visualization in RViz is unclear, rotating the robot at multiple locations may enhance data accuracy and map quality as shown in Figure 3.
e) *Data preservation:* Saving final data with specific names and organized location for easy retrieval and potential future use.

2.2.3 Autonomous radiation mapping. Building upon the manual map acquired previously, the autonomous radiation mapping process is initiated:
a) *Connecting devices:* Connecting the AMoRa to the host computer and the LND7121 GM detector.
b) *Command prompts:* Executing command prompts to launch Python scripts for device connectivity.
c) *Utilizing manual map:* Loading the created 2D map as a reference for autonomous robot navigation.
d) *Grid visualization:* Using RViz to visualize the grid and cell generation, map partitioning, and the robot's path.
e) *Partition verification:* Ensuring accurate partitioning to define the robot's path and prevent overlapping.
f) *GM detector activation:* Establishing a connection to the GM detector, enabling its functionality.
g) *Setting parameters:* Configuring parameters such as grid cell dimensions, counting time for each sampling point (in seconds), and the desired number of partitions.
h) *Autonomous mapping:* Commencing the autonomous mapping process, where the robot follows predetermined paths through the grid to measure radiation levels. If certain spots are unreachable due to obstructions, the system flags them accordingly.

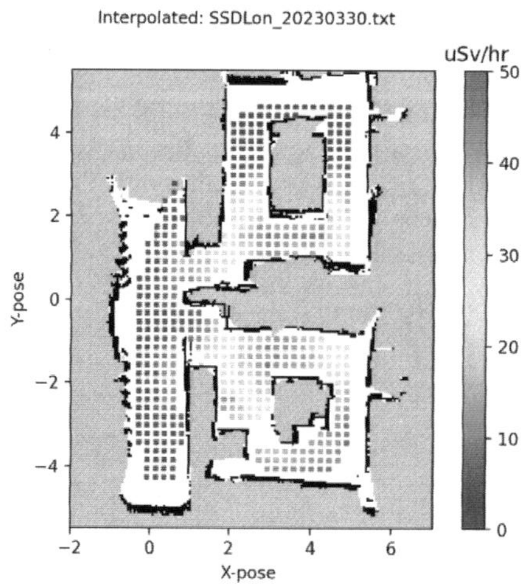

Figure 4: Grid generated and measured by detector

2.2.4. Generation of radiation map and report. Upon completion of the autonomous process, the radiation map is generated based on the visited grid locations, shown in Figure 4. Distinct colour

IOP Conf. Series: Materials Science and Engineering 1308 (2024) 012004 doi:10.1088/1757-899X/1308/1/012004

contours signify radiation intensity, providing a visual representation of hazard levels. The data collected and the results obtained throughout the process are invaluable and must be preserved for future reference.

a) *Report generation:* The generation of comprehensive reports is essential. These reports should be elaborated upon by the operator to provide clear explanations and insights. These reports serve to enrich the data bank and facilitate future mapping needs or emergency responses. For instance, the Malaysian Nuclear Agency (MNA) currently collects data from various sites involving radiation sources. The manual and radiation map data collected are instrumental for present and future requirements, ensuring readiness for any future contingencies or needs.

This revised description employs clearer and more precise terminology to convey the technical aspects of the radiation mapping process with a robot.

3. Results and discussions

As mentioned previously, the final result of the radiation mapping produces a clear picture of intensity of radiation at given grid and also overall color contour on hazard level varies from red to blue color thus determine the spot for source of the radiation. This will help the process of taking the appropriate approach on handling the situation such as calling the authorities.

During this research, there were many challenges and issues faced that can be improved and enhance to get better results and data. One of the main problems faced was the inability of the sensor to detect objects and obstacles that are lower than 20 cm. This is because the position of the RGBD camera is at 20 cm from the ground and it only able to detect on that particular line of height instead of at vertical angle of 43° that the camera capable of which is shown in Figure 5. This is currently being resolved programmatically.

Subsequently, the 2D mapping process of this study needs to be done manually first rather than fully automated. This is important in order to get a more precise data. The robot is able to operate fully automated but it tends randomly miss spots that sometimes crucial for the data, so it is preferred to first run it manually and determine all the grid needed before continuing the automated process.

The issue of terrain limitation is due to the AMoRA small wheels that is only suitable for smooth surface. At some sites even if it is indoor, there will be certain area that has rough and uneven terrain or holes shown in Figure 5. As so temporary solution, thin wooden or metal sheets were used to flatten the surface. The dimension of the AMoRA also restrict the reachability to narrow spots since the size and height that is more than half a meter which resulting in the inability to measure grids that located in those narrow spots that might prove to be important contributor for the data.

Figure 5. Challenges and solutions.

The issue of short life of battery for both robot and host computer that constrained the long operation process. This issue will affect the mapping operation for larger sites that requires longer operation time. In some case where the site consists of symmetrical dimension of site, the sensor tends to overlap the data yielding in error for the data which shown in Figure 3.

Lastly, if the automated mapping is not performed directly after the manual mapping and there is changes in the object's arrangement, the previous data collected during manual mapping might not be able to be used for automated run because of the changes in the environment and leads to the robot to unable to reach the spots that is now blocked by the changes.

Most of the issues and challenges faced for this study can be fixed temporarily or permanently which is by using a better version of robot that have better feature and access such as using robot that operate on robotic, as for this study still in the experimental evaluation phase [4], the current AMoRA proved to be sufficient.

4. Conclusion

In summary, this study presented initial findings from a field test of the AMoRA autonomous radiation mapping robot. It demonstrated the benefits of using autonomous robots for radiation mapping, including improved safety and data accuracy. The methodology involved site surveys, manual 2D mapping, and autonomous radiation mapping. Challenges like object detection limitations and terrain constraints were identified, with temporary solutions implemented. The results highlighted the effectiveness of autonomous radiation mapping, though improvements are needed. This research contributes to the field and sets the stage for future advancements in robotic radiation mapping.

5. Reference

[1] West A, Tsitsimpelis I, Licata M, Jazbec A, Snoj L, Joyce M J, Lennox B 2021 Use of Gaussian process regression for radiation mapping of a nuclear reactor with a mobile robot *Scientific Rep.* **11**(1) 13975.

[2] Grocholsky Ben *et al.* 2006 Cooperative air and ground surveillance." *IEEE Robotics & Automat. Mag.* **13**(3) 16-25.

[3] Grimstad L, From P J 2017 The Thorvald II agricultural robotic system *Robotics* **6**(4) 24.

[4] Abd Rahman N A, Sahari K S M, Jalal M F A, Rahman A A, Abd Adziz M I, Hassan M Z 2020 Mobile robot for radiation mapping in indoor environment *IOP Conf. Ser.: Mater. Sci. and Eng.* **785**(1) 012021 (IOP Publishing).

[5] Gabrlik P, Lazna T 2018 Simulation of gamma radiation mapping using an unmanned aerial system *IFAC-PapersOnLine* **51**(6) 256-262.

IOP Conf. Series: Materials Science and Engineering 1308 (2024) 012005 doi:10.1088/1757-899X/1308/1/012005

Preliminary results of the autonomous radiation mapping in Malaysian Nuclear Agency

Nabilah Ramli[1,a)], Nur Aira Abd Rahman[1], Muhammad Izzuan Mohd Ghazali[1], Suzilawati Muhd Sarowi[1], Muhamad Zahidee Taat[1], Ahmad Hasnulhadi[1], Syirrazie Che Soh[1], Lojius Lombigit[1], Nor Arymaswati Abdullah[1], Noor Farhana Husna A Aziz[1], Mohd Shafiq Sazali[1], Azimawati Ahmad[1], Khairul Salleh Mohamed Sahari[1]

[1]Malaysian Nuclear Agency, 43000, Bangi, Selangor, Malaysia

[a]nabilah@nm.gov.my

Abstract. Autonomous radiation mapping refers to the use of autonomous robots to perform radiation mapping to measure and analyse the levels of radiation intensity of a target area. The primary purpose of the radiation mapping is to uncover the dose distribution across the target area and identify the presence of any hotspots. In this paper, the findings of the autonomous radiation mapping experiments at selected sites in Malaysian Nuclear Agency were presented. The experiments were conducted with a radiation mapping robot namely Autonomous Mobile Robot for Accurate Radiation Detection and Mapping (AMoRA) developed in the previous research. The results consisting of radiation maps of the target areas were presented and discussed.

1. Introduction

Radiation in the workplace originates from various sources and exposure to it can pose health risks to employees if not properly managed. Some common sources of workplace radiation are coming from nuclear reactor and nuclear research facilities, radioactive materials, X-ray machines and radiation therapy equipment. On top of that, there is a regulation that specifies that each licensee and employer engaged in activities associated with regular or potential occupational exposure is required to ensure the safety of their workers from occupational hazards [1]. In order to protect the workers from the harmful effects of radiation, it is imperative to implement safety measures and practices. These measures and practices encompass conducting risk assessments, monitoring radiation levels in the workplace and regularly inspecting and maintaining radiation-emitting equipment to ensure safety [2].

Towards the realisation of this matter, the radiation dose measurement was conducted to measure a radiation dose in the workplace by [3]. The primary objectives of this work include acquiring the radiation dose distribution, creating a dose database and producing a dose map within the facility. The work has been implemented in the waste storage facility using the Ludlum Model 2241-3 survey meter with 118 sampling points were recorded as shown in Figure 1. The process of manually measuring radiation doses for each grid was time consuming and posed a significant safety risk to the workers responsible for this task.

IOP Conf. Series: Materials Science and Engineering 1308 (2024) 012005 doi:10.1088/1757-899X/1308/1/012005

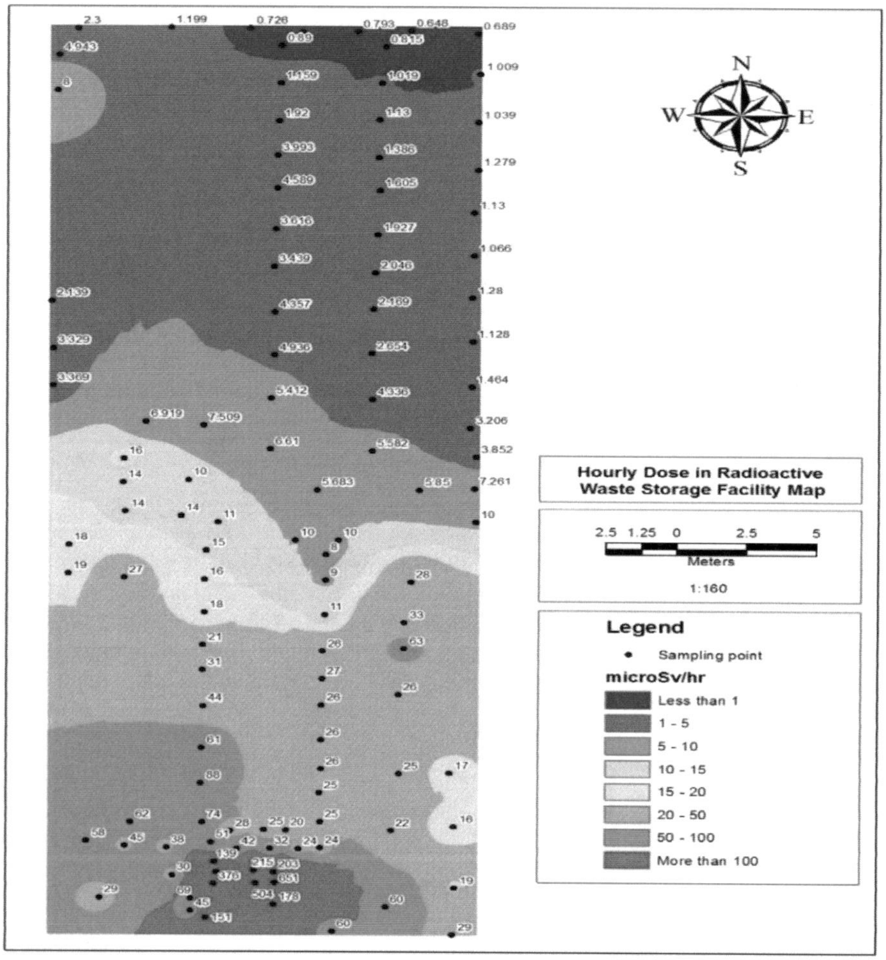

Figure 1. The data grid for the manually recorded radiation readings at the waste storage facility (Harun et. al., 2018)

Due to these circumstances, there is a study from Zakaria et al. (2017) that explores the development of a spatial radiation map by an autonomous mobile robot that is equipped with a Geiger Muller (GM) sensor. The study was intended to develop a gamma radiation mapping system that reads and process location data of a mobile robot with encoder as well as the radiation data transmitted by the GM sensor on the mobile robot [4]. It was conducted with the zigzag waypoints sweeping pattern and the radiation source is placed at a certain point of the designated waypoint. As shown in Figure 2, the result of the experiment conclude that higher radiation activity is recorded when the robot moves near the sources and the radiation activity decreases when the robot .is far from the sources which satisfy the inverse square law theorem.

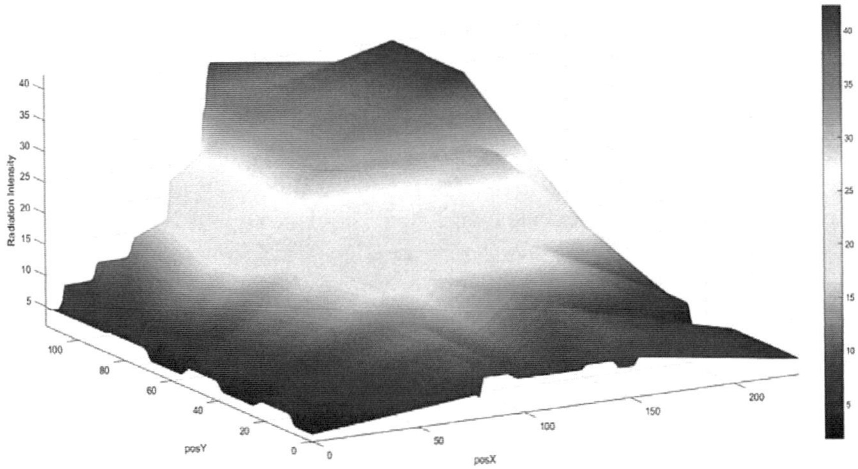

Figure 2. Radiation map with radiation sources [4].

Following that, Autonomous Mobile Robot for Accurate Radiation Detection and Mapping (AMoRA) were developed to map the distribution of radiation in the area [5]. AMoRA ensures measurement accuracy by indicating hotspots on the physical map [6]. This autonomous radiation measurement capability offered by AMoRA can effectively decrease exposure time to radiation. In this study, AMoRA was employed to grid the radiation exposure levels at specific facilities. The facilities were chosen based on the operational activities and associated risk factors.

The primary goal of this paper is to observe AMoRA capabilities in producing the radiation mapping autonomously. As well as to assess radiation distribution, enabling the identification of areas with higher radiation levels to be avoided during work at these facilities. This approach ensures that work activities in the selected areas can be more efficiently coordinated in accordance with the ALARA (As Low As Reasonably Achievable) principle.

2. Radiation mapping

AMoRA consists of a Turtlebot2 as a mobile robot platform together with a GM detector LND7121 as the radiation detector [5]. The GM electronic module is connected to the Turtlebot2 host computer via USB connection as shown in Figure 3.

Figure 3. Complete set of AMoRA used in this study.

Initially, the sites survey was performed to assess the suitability and necessity for the autonomous radiation mapping for a number of potential sites. It was conducted by the team members including radiation safety personnel along with the site owner. Sites survey was done by taking into account a few criteria such as train and surface conditions. This is because of the limitations of the robot which can only move on flat and smooth surfaces. Based on the owner and radiation safety personnel, a necessity of performing the operation will be analysed and decided depending on the operating conditions the sources involve.

Next, a 2D map of each selected site was created with Simultaneous Localisation and Mapping (SLAM) by running the Robot Operating System (ROS) *gmapping* package. The procedure was manually operated by personnel and created a 2D map. Both processes are presented in Figure 4(a) and 4(b) respectively.

(a)

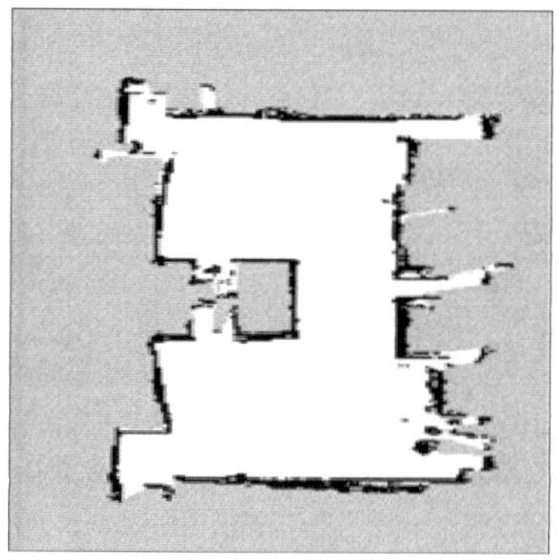

(b)

Figure 4. a) The AMoRA was manually deployed to create a 2D mapping operation, b) the generated 2D map of the area.

The following step is shown in Figure 5, where AMoRA is deployed to perform autonomous radiation mapping by executing the ROS *rad_mapper* package developed in the previous work [3]. The *rad_mapper* divides the 2D map into a grid and generates sampling points at the centroid of each unoccupied grid cell shown in Figure 6. The counting time for data collection can be configured manually. The mapping may be repeated at different conditions depending on the requirement of the site and will be discuss in the discussion section. Finally, the radiation map was generated and a corresponding report was produced. However, in the cases where the site is found to be unsuitable during the site visit, a report is generated to explain the reasons for its unsuitability.

Figure 5: AMoRA in autonomous mode mapping the area

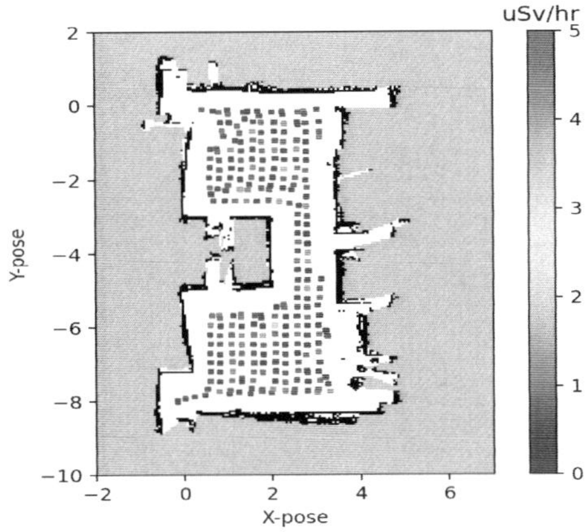

Figure 6: Robot trajectory during the autonomous radiation mapping

3. Results and discussion

The experiment was carried out at the facilities of the Malaysian Nuclear Agency under the guidance of professional supervision. There are three of the results presented in this work which are BTP Development Laboratory, Interim Storage Facility and Reactor TRIGA PUSPATI. This section discussed the finding and results of the experiment. The scale indicator of each location varies depending on their radiation distribution to indicate the hotspot of the specific area. The colour contour of the indicator shows the difference in radiation level. Starting from blue colour indicate the background reading, while the colour gradient gradually ascends from blue to red colour indicated the increment of hazard level. Hence, the red colour in the map signifies the location have the high intensity of radiation.

3.1 BTP Development Laboratory

The radiation mapping at BTP Development Laboratory was executed twice. Initially during the rest state, where the radioactive material source remained in the generator with proper shielding. The result shown in Figure 7(a) demonstrate only background reading was detected at rest state. After that, AMoRA was deployed again during the workload state. The workload state is the process where the radioactive source (Ga-68/Tc-99m) which is eluted from the generator and transferred into a shielded fume hood for research work. During the ongoing work, there is a concern from the employer regarding the effectiveness of the shielding in preventing radiation exposure to personnel. The results obtained clearly demonstrate the presence of radiation distribution during the work as shown in Figure 7 (b). It proves that the radiation around the working area is slightly elevated.

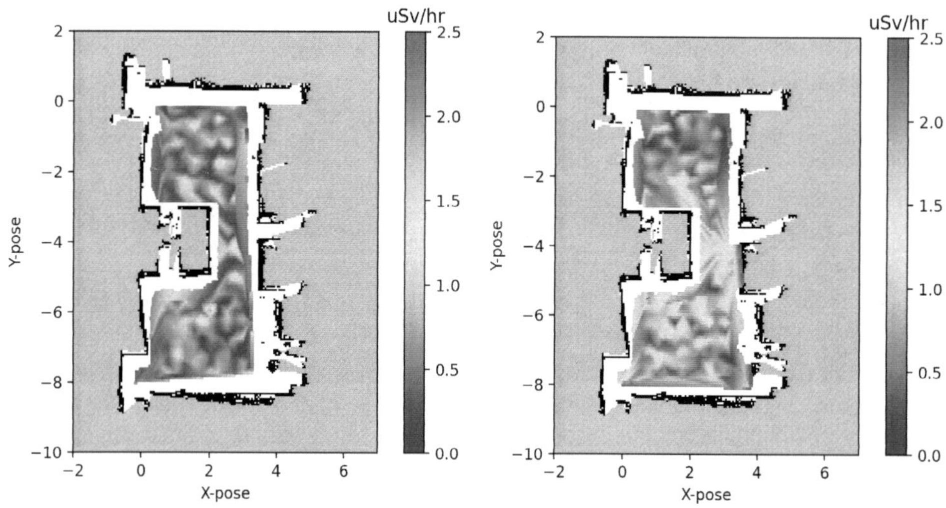

(a) (b)

Figure 7. Radiation mapping contour map at BTP Development Laboratory, a) at rest, b) at workload.

Finding of this is useful for facility personnel to evaluate and coordinate the activities to ensure that the radiation exposure is within permissible limits. Based on the findings provided, employer can strategize to enhance shielding measures and restructure their work procedures to minimise radiation exposure.

3.2 Interim Storage Facility

The Interim Storage Facility serves as a storage site for low-level radioactive material originating from reactor operations, medical, academic, industrial and other commercial uses of radioactive materials [7]. The configuration of radioactive material within this facility is constantly changing. Radiation mapping is carried out to observe scattered radiation and detect areas with elevated levels of radiation. The results obtained clearly reveal hotspots in various areas within the facility. This contour map allows for the precise identification of hotspot regions because the mapping is conducted on the actual location map as can be seen in Figure 8. In contrast, manual mapping [3] only offers a general location estimate for the hotspot. Additionally, the exposure to radiation workers during the radiation mapping process is minimised by the autonomous operation.

IOP Conf. Series: Materials Science and Engineering 1308 (2024) 012005 doi:10.1088/1757-899X/1308/1/012005

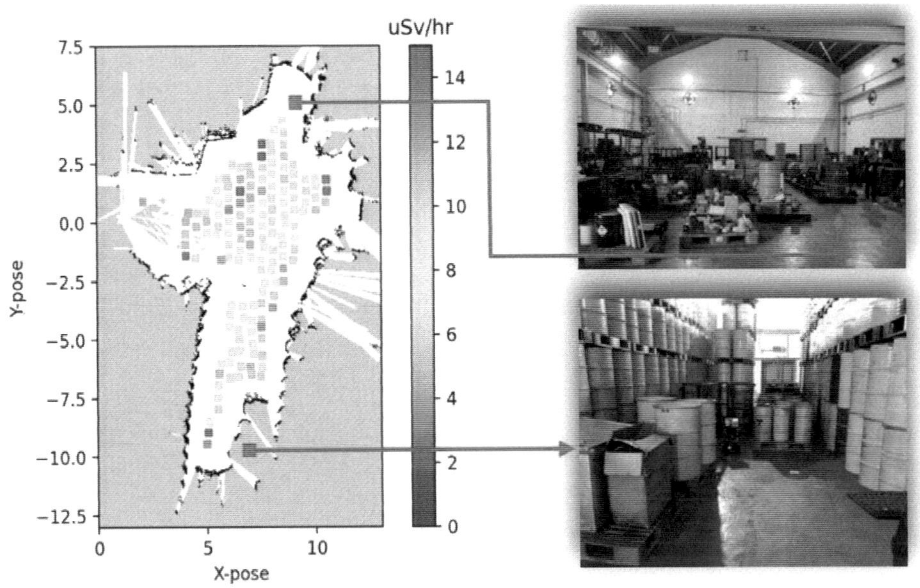

Figure 8. Hotspot areas are cleary visible in several location.

3.3 Reactor TRIGA PUSPATI

Research Reactor TRIGA PUSPATI equipped with the facilities such as Neutron Radiography (NUR), Small Angle Neutron Scattering Neutron (SANS), Diffraction and Thermal Column for Boron Neutron Capture Therapy (BNCT). Research activities are conducted at each of these facilities while the reactor is in operation. The radiation levels inside the reactor hall were monitored by area monitoring instruments in each corner of the building. However, despite all the facilities having shielding in place, scattering radiation remains a concern. To provide a detailed view of radiation distribution inside the reactor hall, autonomous radiation mapping was carried out. As anticipated, the highest radiation hotspots are observed around facilities like between Diffraction and Thermal Column for BNCT as shown in Figure 9. This contour map precisely pinpoints the location of these hotspots.

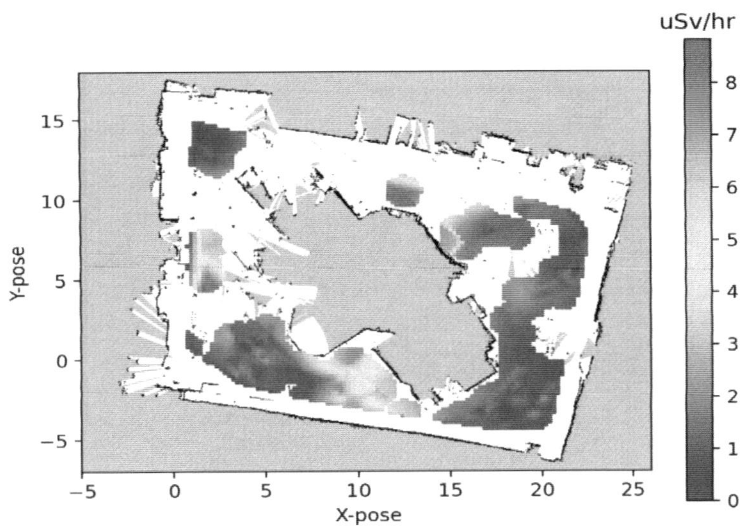

Figure 9. Distribution of the radiation inside the reactor hall.

4. Conclusions and future work

As a conclusion, the results presented in this paper have demonstrated AMoRA capabilities to perform autonomous radiation mapping and produce comprehensive radiation maps of the target areas. These radiation maps can be referred to effectively manage and minimise radiation exposure in the workplace, to ensure the safety compliance with local regulations and standards. Based on issues and challenges observed during AMORA deployment at the selected site, the design and algorithm will be improved in the future.

5. References

[1] Act A E L 2010 Atomic energy licensing (Basic safety radiation protection) regulations 2010.

[2] Cuaron J J, Hirsch A E, Medich D C, Hirsch J A, Rosenstein B S 2011 Introduction to radiation safety and monitoring *J. of the American College of Radiol*.**8**(4) 259-264.

[3] Harun N, Sujan M F, Ibrahim M Z 2018 Spatial interpolation of gamma dose in radioactive waste storage facility *IOP Conf. Ser.: Mater. Sci. and Eng.* **298**(1) 012047 (IOP Publishing).

[4] Zakaria A H, Mustafah Y M, Abdullah J, Khair N Abdullah T 2017 Development of autonomous radiation mapping robot *Procedia Comp. Sci.* **105** 81-86.

[5] Abd Rahman N. A, Sahari K S M, Jalal M F A, Rahman A A, Abd Adziz M I, Hassan M Z 2020 Mobile robot for radiation mapping in indoor environment *IOP Conf. Ser.: Mater. Sci. and Eng.*.**785**(1) 012021(IOP Publishing).

[6] Abd Rahman N A, Sahari K S M, Hamid N A, Hou Y C 2022 A coverage path planning approach for autonomous radiation mapping with a mobile robot. International *J. of Adv. Robotic Sys.* **19**(4) 17298806221116483.

[7] US NRC 2015 Backgrounder on radioactive waste https://www.nrc.gov/reading-rm/doc-collections/factsheets/radwaste.html.

Acknowledgement

Special thanks to En Wan Hamirul Bahrin from BTP Development Laboratory, En Ahmad Hasnulhadi from Interim Storage Facility, En Na'im Syauqi and Mohamad Amirudin from Reactor TRIGA PUSPATI for granting us permission to conduct our work at respected facilities, providing valuable input and feedback.

IOP Conf. Series: Materials Science and Engineering 1308 (2024) 012006 doi:10.1088/1757-899X/1308/1/012006

Lagrangian and Eulerian approach to predict the movement of radionuclides in selected potential sites in Malaysia during the monsoon period

L H Muhamad[1,a)], M K A Karim[1], K A Yusof[1] and N A Basri[2]

[1]Department of Physics, Faculty of Science, Universiti Putra Malaysia, 43400, Serdang, Selangor, Malaysia
[2]Department of Energy Engineering, Faculty of Engineering, Universiti Teknologi Malaysia, Skudai, Malaysia

[a]lokmanhaakimm@gmail.com

Abstract. In computing the movement of particles through the fluid, three types of dispersion models are used which are Gaussian plume models, Lagrangian puff models, and small-scale numerical models. Lagrangian particle dispersion models are increasingly used for nuclear applications. In this study, the usage of the Lagrangian model is implemented in a Hybrid Single Particle Lagrangian Integrated Trajectory (HYSPLIT) which combines with the Eulerian model to predict the movement of radionuclides from the simulated nuclear power plant accident in selected locations in Malaysia. These include nuclear risk studies in these areas, emergency response systems, and source term analyses. This study aims to compare the risk from the dispersion of ^{137}Cs from a simulated Nuclear Power Plant (NPP) at Mukah and Tasik Kenyir during the Northeast monsoon (8th October 2022) and Southeast monsoon (16th May 2022). The HYSPLIT model was set up based on the meteorological data during both monsoon periods and was simulated for 5 days. The outcomes revealed that the concentration of ^{137}Cs is higher at Tasik Kenyir compared to Mukah for both during the Northeast monsoon period (3.5 x 10^7 Bq/m^2) while for Southwest monsoon, the result is vice versa where the concentration of ^{137}Cs at Mukah is (2.2 x 10^7 Bq/m^2). The activity concentration is then converted into rate effective dose measurement and compared with the limit suggested by ICRP which is 1 Sv per year. It is concluded that Tasik Kenyir is a more suitable location to build NPP compared to Mukah as the movement of radionuclides was dispersed to the location that contained less population.

1. Introduction

The world's population has been growing at a pace rarely seen before, and this demographic shift has a big impact on how much energy the world needs. According to the United Nations [1], the world's population has increased to over 8 billion in 2023. This increase, together with industrialization and urbanization, has resulted in a rising demand for dependable and sustainable energy sources to power our contemporary society. In Southeast Asia, the absence of nuclear power plants is notable; however, as the population continues to grow and energy demand escalates, it becomes imperative to proactively identify potential sites for such facilities. With energy demand in Malaysia steadily rising at an annual rate of approximately 3-4%, it is evident that the country urgently requires a nuclear power plant to address its growing power needs effectively. However, due to the harmful effects that have resulted from previous nuclear accidents (e.g., Fukushima and Chernobyl), many people are apprehensive about the use of nuclear energy. In nuclear safety, it is important to predict the movement of radionuclides from the source to protect humans from being exposed to them. Many types of atmospheric models can be used to predict the movement of particles such as AERMOD, HOTSPOT, CALPUFF, and HYSPLIT.

HYSPLIT is an atmospheric model developed by the National Oceanic and Atmospheric Administration (NOAA) in 1980. The model calculation in HYSPLIT is a hybrid model between the Lagrangian-Eulerian System. The HYSPLIT model is made to support a variety of simulations including the atmospheric transport and dispersion of pollutants and hazardous materials as well as the deposition of these materials to the Earth's surface. Tracking and predicting the release of radioactive material, volcanic ash, wildfire smoke, dust, and pollutants from different fixed and mobile emission sources are some of the usages [2]. HYSPLIT Model of the Air Resources Laboratory of NOAA database was also used along with isotopic compositions to study atmospheric moisture by measuring the concentration of Oxygen-18 and Hydrogen-2 in the Bojnourd area, Iran [3]. Other than that, in the nuclear safety sector, a study has been made to analyze the radiological component on a potential site at Baiji. The results of the effective dose were compared with the dose limits that are suggested by the International Commission on Radiological Protection (ICRP), and it exceeded by 1 Sv for the areas that are near the potential site [4]. Thus, in this study, nuclear incidents were simulated for some potential NPP sites in Malaysia during critical periods in the Northeast Monsoon and Southwest monsoon.

2. Study Area

Tasik Kenyir (2.8921° N, 112.0922° E) and Mukah (5.0040° N, 102.6388° E) are the selected NPP potential sites in Malaysia. Both locations are located near the sea and the population density is less than the limit set by AELB which is 200 people/m2 [5]. The area surface of Tasik Kenyir and Mukah is about 270 km^2 and 2536 km^2 respectively. Previously, there was a study about the site selection in Malaysia in Johor State, where Tenggaroh and Jemaluang are the most suitable locations among the other sites potential. Based on the AELB procedure, they outline a three-step process to determine the location of the NPP [6]:

i. It cannot be affected by phenomena against which protection through the design is impracticable.
ii. The probability of occurrence and the severity of destructive phenomena are not too high.
iii. The site characteristics are such that the consequences of a potential accident would be at acceptable limits.

3. Methodology

In predicting the movement of particles in the atmosphere, the Lagrangian model is one of the atmospheric models that is usually used in many fields and it is also considered as one of the emergency preparedness tools. The Lagrangian technique predicts the movement and dispersion of pollutants by following the trajectories of a large number of "particles" that originate from a source point. Equation 1 shows the calculation of x-y components for a 2-dimensional grid combined with equation 2 which calculates the altitude, and z of the particle by considering the average turbulence velocity of the wind [7].

$$X_{x,y}\ t + \Delta t = X_{x,y}\, t + \ \bar{U}_{u,v,w}\, t + U'_{u,v,w}\ t\ \Delta t \quad (1)$$
$$)$$

$$Z_{\text{final}}(\ t + \Delta t) = Z_{\text{mean}}(\ t + \Delta t)\ +\ W'(\ t + \Delta t)\ \Delta t\ Z_{\text{top}}^{-1} \quad (2)$$

There are two ways to calculate particles: vertically or simultaneously in both vertical and horizontal directions. There are two possible horizontal distributions for calculations with a vertical particle distribution: top-hat and Gaussian puff. On the other hand, instead of being calculated at a grid point, particle computations are added to a grid cell. The area of a cell, which corresponds to the distance between adjacent nodes in half, is specified at the node's center. One particle of mass m is defined as contributing an incremental amount of concentration to a cell as a 3D particle. The concentration of pollutant's mass at a specific grid inside the atmosphere is calculated as shown in equation 3 [7].

$$\Delta c = m(\Delta x\ \Delta y\ \Delta z)^{-1} \quad (3)$$

On the other hand, the Eulerian model is an approach that specifies a specific grid to calculate the velocity and concentration of particles concerning space and time, The mass balance in the Eulerian model is centered on a differential volume. The air motion is computed using the velocity field [8]. Figure 1 shows the fixed three-dimensional grid which acts as a frame of reference to compute the pollutant air concentrations.

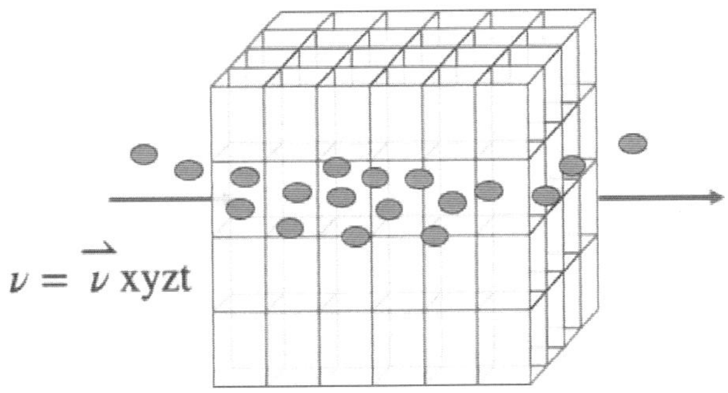

Figure 1. Movement of particles through the specific grid in the Eulerian approach by respect to time and space,

In this study, HYSPLIT software is used to simulate the movement of radioactive particles in the environment utilizing atmospheric dispersion modeling. The National Oceanic and Atmospheric Administration (NOAA) of the USA created the computer model HYSPLIT to measure the paths of air parcels and the concentration of air contaminants. As mentioned earlier, this software applies both Lagrangian and Eulerian approaches to simulate the movement of particles in an interval period. HYSPLIT requires the input of meteorological data and usually the input is obtained from The National Centers for Environmental Prediction (NCEP) and the National Center for Atmospheric Research (NCAR) of the USA supply open access [2] via (https://www.ready.noaa.gov/data/archives/gdas1). Meteorological data that is inserted in this model is called a Global Data Assimilation System (GDAS) which is considered global meteorological data that contains variables such as temperature, pressure, relative humidity, and wind direction. This parameter is simulated by HYSPLIT to produce a trajectory that might travel by radionuclides in the simulation.

By using meteorological data from 2022 in May and October, the trajectory simulation was done for a daily emission throughout 5 days. The emission input is designed to be equal to the predicted release of cesium during the Fukushima accident in 2011 to obtain a clear observation of air movement in an intense nuclear emission scenario [9]. As the emission origin, the emission height was chosen at 0 meters above sea level as it is a significant height for humans in that area. After computing the simulation, the activity concentration for both locations in both periods, the activity concentration is converted into rate effective dose rate by using the dose rate coefficient for ^{137}Cs and Equation 3.

$$E = e_E At \quad (4)$$

where e_E is the dose rate coefficient, A is the activity concentration and t is the duration of a person being exposed to the pollutant. ^{137}Cs are chosen for this simulation because it has a long half-life of 30 years and can emit gamma radiation. Since ^{137}Cs stay longer in the environment, they can travel a long distance across the globe. Long-term exposure to Caesium can cause stochastic health effects such as cancer or tumors [10]. Table 1 shows the radionuclide input data and the selected period in this simulation.

Table 1. Input parameter for a hypothetical accident using HYSPLIT

Information	Northeast Monsoon	Southeast Monsoon
Starting time	8th October 2022	16th May 2022
End time	12th October 2022	20th May 2022
Phases	Particle	
Radionuclide	^{137}Cs	
Half-life (days)	10,960	
Emission Rate	1.53 X 10^{16} Bq/event	
Release start	0000 UTC	
Altitude	0 m	

4. Result and discussion

4.1. Analyze meteorological data

Based on the GDAS meteorological data, parameters such as temperature, wind speed, and wind direction were extracted and then compared within both monsoons. The interval time for these parameters' measurement is 3 hours. Based on figure 2(a), shows that the temperature at Tasik Kenyir has a higher range compared to the Mukah area where the differences in the temperature during the day and night are large for Tasik Kenyir. The Southwest monsoon (Figure 2(b)) follows a similar pattern for both sites, but Tasik Kenyir's maximum temperature is greater than the Northeast monsoon's, which is roughly 34.3°C.

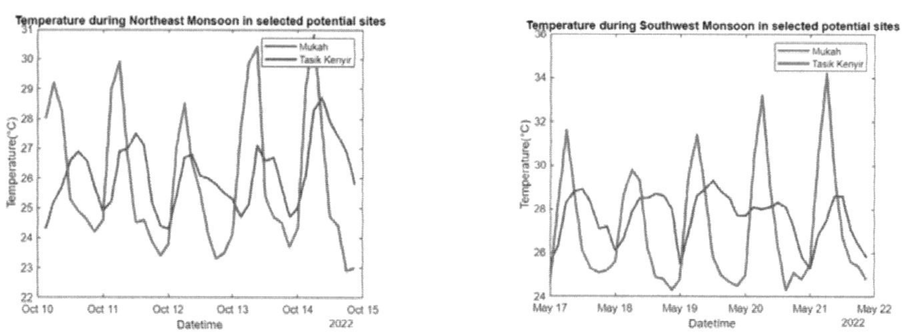

(a) (b)

Figure 2. (a) Temperature for both locations during the Northeast monsoon and (b). during the Southwest monsoon.

To visualize the wind data before converting it into trajectory form, the wind speed and wind direction are then plotted in a wind rose diagram (figure 3(a) and 3(b)). According to the data, the wind direction for Mukah during the Southwest monsoon moved in a wide range of directions and did have an impact on the dispersion of radionuclides, which dispersed into plenty of locations. The same goes for the movement of radionuclides at Tasik Kenyir during the Northeast monsoon where it dispersed in many directions. Similarly, radionuclides moved over from Tasik Kenyir during the northeast monsoon and scattered in all directions.

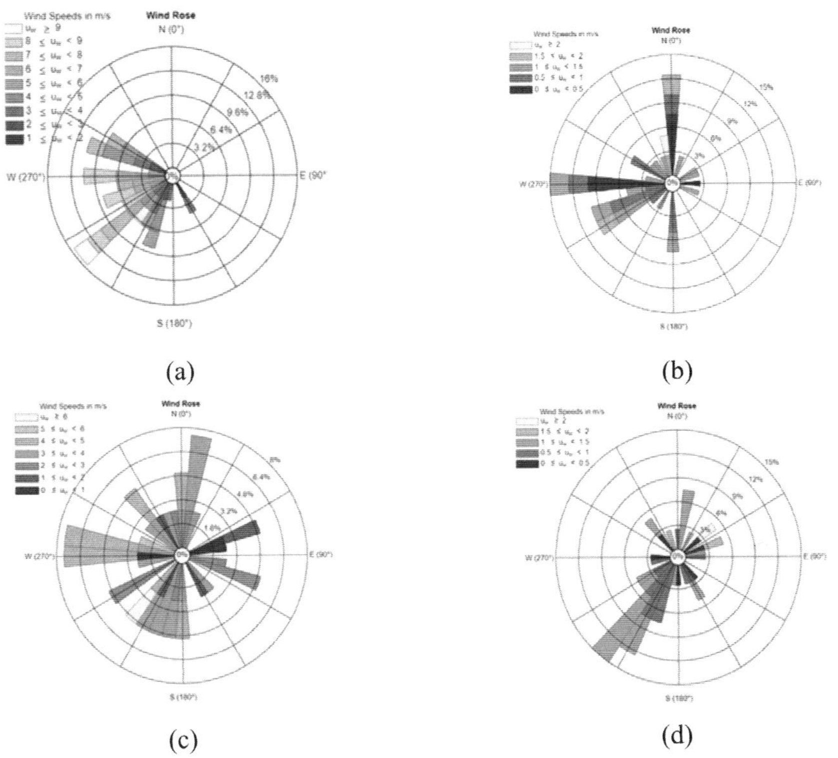

Figure 3. Wind rose at (a)Mukah, Northeast monsoon. (b) Tasik Kenyir, Northeast monsoon. (c) Mukah, Southwest monsoon, and (d) Tasik Kenyir, Southwest monsoon

4.2. Trajectory analysis

Wind trajectory analysis is a valuable tool for understanding the movement of air masses and their potential impacts on our environment and industries. By tracking wind patterns and predicting their behavior, it can be used as an initial risk assessment right after the onset of a nuclear accident. Immediate emergency evacuation or sheltering orders can be carried out through this analysis, thus minimizing the affected population's exposure to radioactive deposition. Based on figure 4(a), shows the wind direction moving to the north-east direction for both locations. However, during the southwest monsoon period in Mukah (Figure 4(c)), the wind direction for Mukah was predicted to move and accumulate in the region that has a high population in Sarawak state.

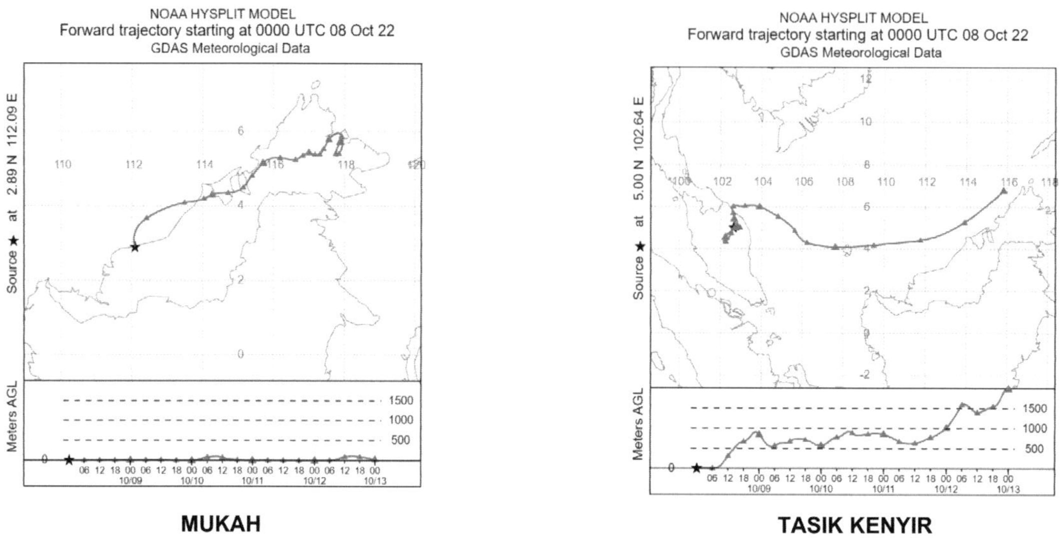

(a)

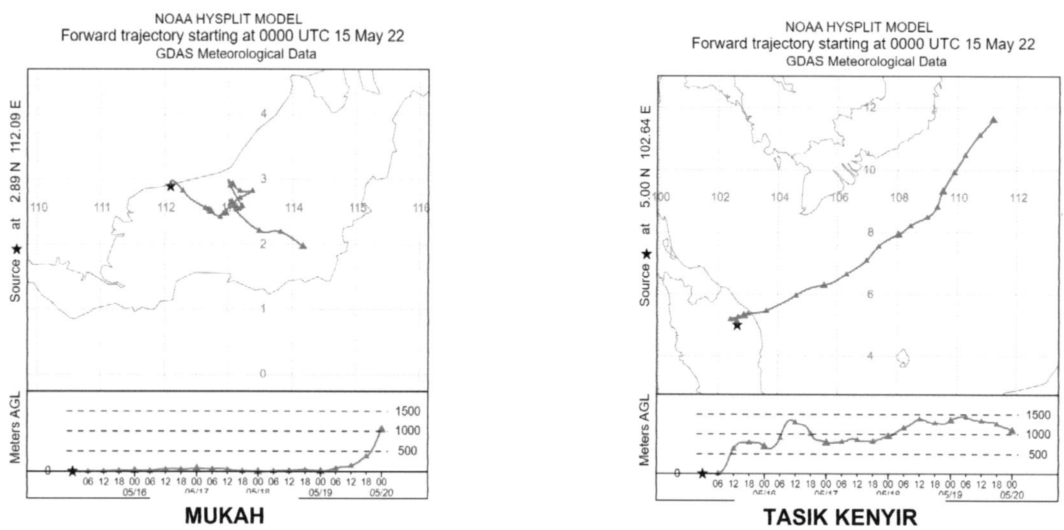

(b)

Figure 4. Trajectory analysis in the Tasik Kenyir and Mukah region during (a) Northeast and (b) Southwest monsoon.

4.3 Activity concentration analysis

The dispersion of ^{137}Cs has been evaluated for the first 5 days after the hypothetical accident starting at UTC 00:00 at the ground level. The main purpose of the 0 m altitude is to evaluate the value that is most exposed to human beings. To give a route for escape in the event of an unexpected catastrophe, this prediction should be helpful in any emergency preparedness plans. Based on Table 2 and Table 4, the dispersion of ^{137}Cs was estimated to reach Tanjung Beringin, West Kalimantan, Indonesia after 5 days of a nuclear accident. The movement of radionuclides in Mukah has affected the region that contains a large population. For Tasik Kenyir during the northeast monsoon, the movement of radionuclides reached some regions in Sabah state on the 5th day but the concentration was not as high as in Mukah. During the southwest monsoon, the movement of ^{137}Cs has reached Peninsular Malaysia from Mukah. It is because of the wind direction that spread to various directions as observed in Figure 3 (c).

The dose rate coefficient for ^{137}Cs at 0 m for newborns, teenagers, and adult persons are 9.23 x 10-18 Sv Bq-1 s -1 m2, 7. x 10-18 Sv Bq-1 s -1 m2 and 7.85 x 10-18 Sv Bq-1 s -1 m2 respectively [11]. This dose rate coefficient is used to convert activity concentration into effective dose rate measurement. This study focuses on the dose rate coefficient for adult people and the highest amount of effective dose are tabulated in Tables 4 and 5.

Table 2. Atmospheric dispersion of ^{137}Cs during Northeast monsoon.

Location	Activity Concentration (Bq/m^2)	
	1 day	**5 day**
Mukah	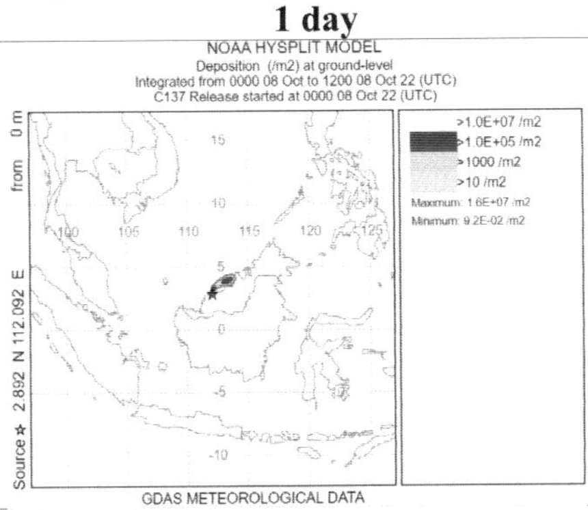	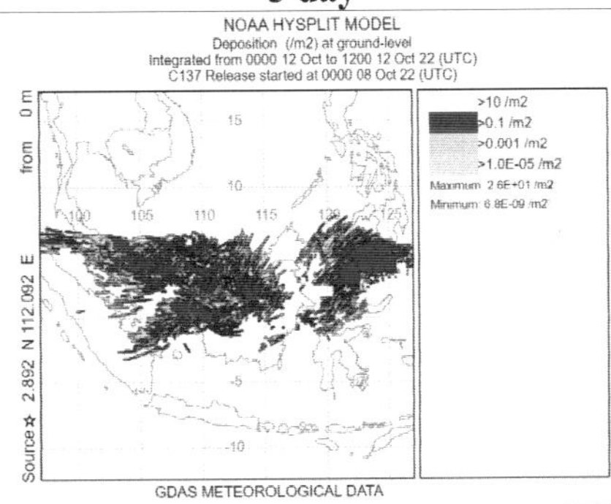
Tasik Kenyir	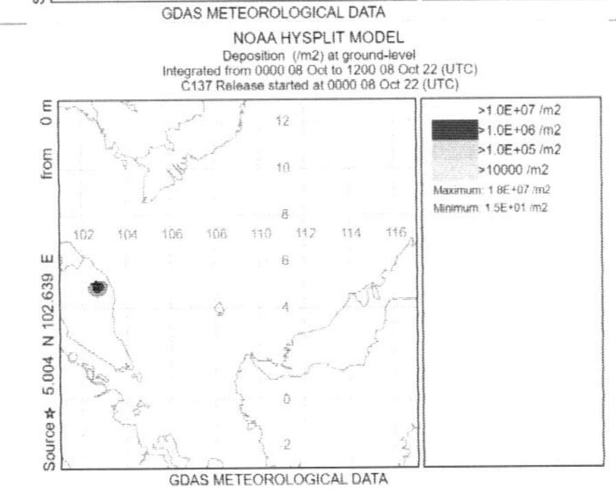	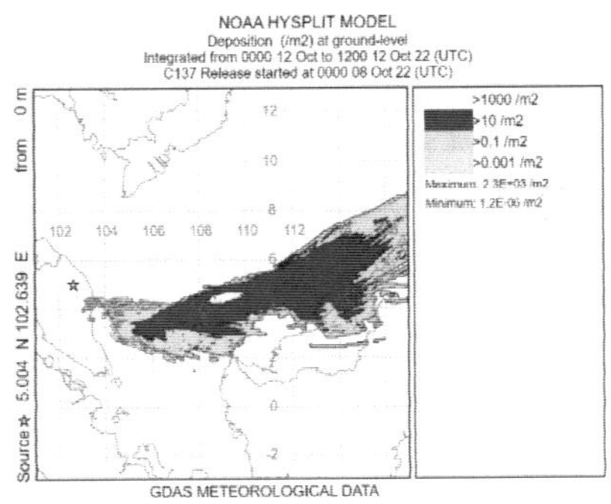

IOP Conf. Series: Materials Science and Engineering 1308 (2024) 012006 doi:10.1088/1757-899X/1308/1/012006

Table 3. Atmospheric dispersion of ^{137}Cs during Southwest monsoon.

Location	Activity Concentration (Bq/m^2)	
	1 day	**5 day**
Mukah	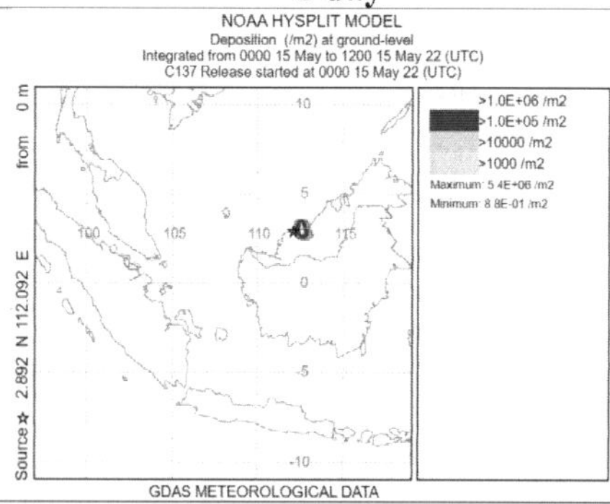	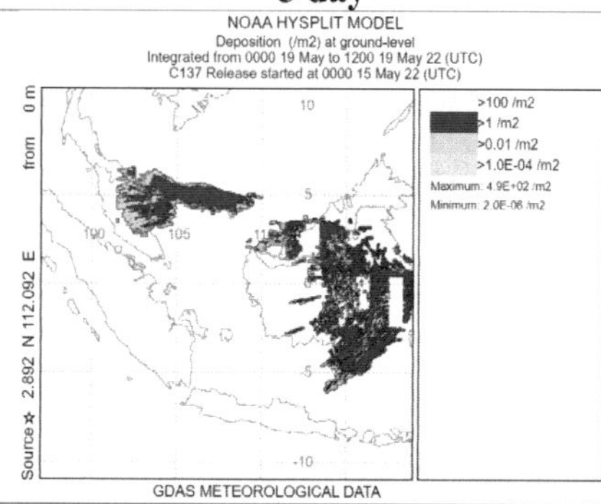
Tasik Kenyir	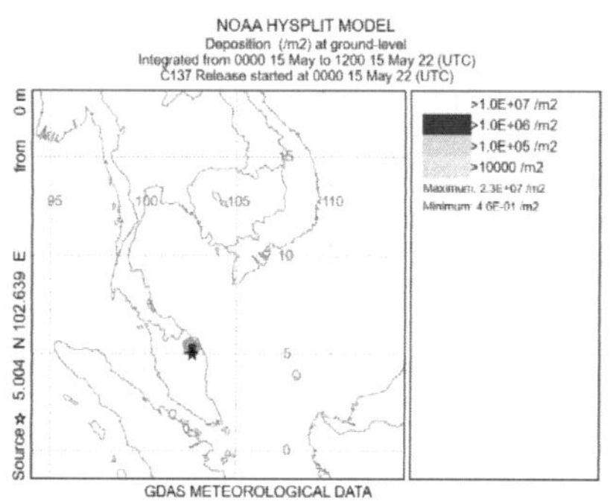	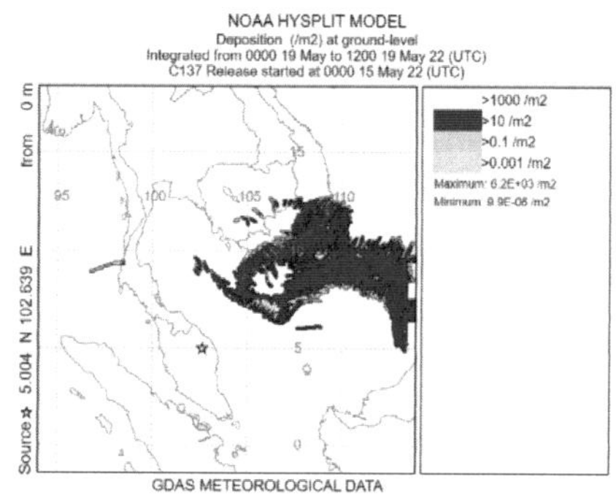

Table 4. Highest activity concentration and highest rate of effective dose during Northeast monsoon.

Location	Number of days	Highest Activity Concentration (Bq/m^2)	Highest Rate Effective Dose (mSv / s m^2)	Coordinates	Location affected	The number of the population affected
Mukah	1	18000000	1.41 x 10^{-7}	2.99 ° N 112.54 ° E	Mukah, Sarawak	42,275
	5	44	3.45 x 10^{-13}	0.66 ° S 109.59 ° E	Tanjung Beringin, West Kalimantan, Indonesia	38,864
Tasik Kenyir	1	35000000	2.75 x 10^{-7}	4.85 ° N 102.59 ° E	Gua Musang, Kelantan	<1000
	5	220	1.73 x 10^{-12}	4.95 ° N 113.59° E	Sea	-

IOP Conf. Series: Materials Science and Engineering 1308 (2024) 012006 doi:10.1088/1757-899X/1308/1/012006

Table 5. Highest activity concentration and highest rate of effective dose during Southwest monsoon.

Location	Number of days	Highest Activity Concentration (Bq/m^2)	Highest Effective Dose (mSv / s m^2)	Coordinates	Location affected	The number of the population affected
Mukah	1	22000000	1.727×10^{-7}	2.74 ° N 112.29° E	Mukah, Sarawak	42,275
	5	320	2.512×10^{-12}	-2.16 ° N 114.54° E	Central Kalimantan, Indonesia	2,650,000
Tasik Kenyir	1	19000000	1.4915×10^{-7}	5.00 ° N 102.69° E	Kampong Buai, Terengganu	<1000
	5	27000	2.1195×10^{-10}	7.10 ° N 113.54° E	Sea	-

5. Conclusion

Based on this study, it is found that Tasik Kenyir is the more suitable location to build NPP as the dispersion of radionuclides does not affect a large number of the population compared to Mukah. However, it is important to provide an emergency plan for the population to evacuate if a nuclear accident occurs. Thus, several atmospheric models such as the Lagrangian and Eulerian models might be useful for making a prediction movement of radionuclides in a certain period.

6. References

[1] Nations U 2023 *World Population Day | United Nations* (United Nations) https://www.un.org/en/observances/world-population-day#:~:text=State%20of%20World%20Population%202023%3A%208%20Billion%20Lives%2C%20Infinite%20Possibilities

[2] Rolph G D, Stein A, Barbara 2017 Real-time environmental applications and display system: READY *Environ. Modelling and Software* **95** 210–228. https://doi.org/10.1016/j.envsoft.2017.06.025

[3] Bagheri R, Bagheri F, Karami G H, & Jafari, H. (2019). Chemo-isotopes (18O & 2H) signatures and HYSPLIT model application: Clues to the atmospheric moisture and air mass origins. *Atmospheric Environment*, *215*(August), 116892. https://doi.org/10.1016/j.atmosenv.2019.116892 5]

[4] Mohammed Saeed, I. M., Mohammed Saleh M A, Hashim S, bin Ramli A T, Al-Shatri S H H 2019 Atmospheric dispersion modeling and radiological safety assessment for expected operation of Baiji nuclear power plant potential site. *Annals of Nuclear Energy*, *127*, 156–164. https://doi.org/10.1016/j.anucene.2018.11.045

[5] AELB 2011 *Guideline for site selection for nuclear power plant.*

[6] Basri N A, Ramli A T 2012 Selection of possible candidate area for nuclear power plant in Johor, Malaysia *J. Nucl. Relat. Technol.* **9** 56–63.

[7] Draxler R R, Hess G D 1998 An overview of the HYSPLIT_4 modeling system for trajectories, dispersion, and deposition. *Australian Meteorol. Mag.* **47**(4), 96 295–308 https://www.arl.noaa.gov/documents/reports/MetMag.pdf

[8] Leelőssy Á, Lagzi I, Kovács A, Mészáros R 2018 A review of numerical models to predict the atmospheric dispersion of radionuclides *J. of Environ. Radioact.* **182** 20–33https://doi.org/10.1016/j.jenvrad.2017.11.009.

[9] Rosen A 2012 Analysis of WHO report on Fukushima catastrophe.

[10] Walker, S. (2002). EPA Facts About Cesium-137. 99(July), 2.

[11] Bellamy, M.B., Dewji, S.A., Leggett, M.R., Hiller, W., Veinot, K., Manger, R.P., Eckerman, K.F., Ryman, J.C., Easterly, C.E., Hertel, N.E., Stewart, D.J., 2019. Federal Guidance Report No. 15 External Exposure to Radionuclides in Air, Water and Soil. External Dose Rate Coefficients for General Application. U.S. Environmental Protection Agency, Washington. EPA 402-R-19-002.

Acknowledgment
This research was funded by Universiti Putra Malaysia under grant vot GP/UPM-GPB-9704400.

Design and integration of a radiation detector module for robot operating system (ROS)

Nur Aira Abd Rahman[1,2,a)], Khairul Salleh Mohamed Sahari[1], Muhammad Zahidee Taat[2], Lojius Lombigit[2], Nor Arymaswati Abdullah[2], NFH Aziz[2], N Ramli[2], Syirrazie Che Soh[2], Muhammad Izzuan Mohd Ghazali[2], MZ Baharuddin[1], Leong Yeng Weng[1]

[1]Universiti Tenaga Nasional (UNITEN), 43000 Kajang, Selangor, Malaysia
[2]Malaysian Nuclear Agency, 43000, Bangi, Selangor, Malaysia.

[a]nur_aira@nm.gov.my

Abstract. In this paper, we present a radiation detector module that can be seamlessly integrated with Robot Operating System (ROS) to enable robots to perform radiation measurements in hazardous environments. The module is designed with a detector PCB and connectors that are compatible with an Arduino shield. The Arduino firmware is programmed with a counter-timer algorithm and publishes data to the ROS environment, allowing for easy visualization of the data in a 2D occupancy map. Our experimental results demonstrate the module's effectiveness in inspecting and reconstructing the robot's path during operations. This paper provides a valuable contribution to the field of robotics by enabling robots to perform radiation measurements safely and accurately in dangerous environments.

1. Introduction

In nuclear and radiation industries, robots can help reduce the risk of radiation exposure to workers [1]–[3]. Robots can perform tasks that are tough or unsafe for humans. Marques et al. [4] identified four key situations where mobile radiation detection and measurement using robots is crucial: radiological and nuclear emergencies, preventing illegal activities, ensuring safety at nuclear facilities, and monitoring naturally occurring radioactive materials (NORM).

To make a robot capable of radiation inspection and monitoring, robot integrated with a radiation detector module is required. Nowadays, many robots are equipped to run on Robot Operating System (ROS), which is open-source software with extensive libraries and tools for building robotics applications [5]. Using ROS makes it faster to develop and test algorithms. It also offers diagnostic and visualization tools like Gazebo and RViz for evaluating algorithms in simulations and real-world experiments [6].

This paper introduces the design and integration of a radiation detector module that seamlessly works with ROS. This module serves as a fundamental building block for robots operating in radiation environments and can be adapted for various applications, including surveillance, emergency response, and monitoring. Additionally, it can be a valuable resource for students learning ROS and working on robotics projects.

2. Methodology

2.1. Overview of the system

In this section the design and integration of the GM module with ROS will be described. The system comprises two main components: the GM module and a host computer running ROS. Figure 1 illustrates the block diagram of this system, while Figure 2 provides a visual representation of the physical connections involved.

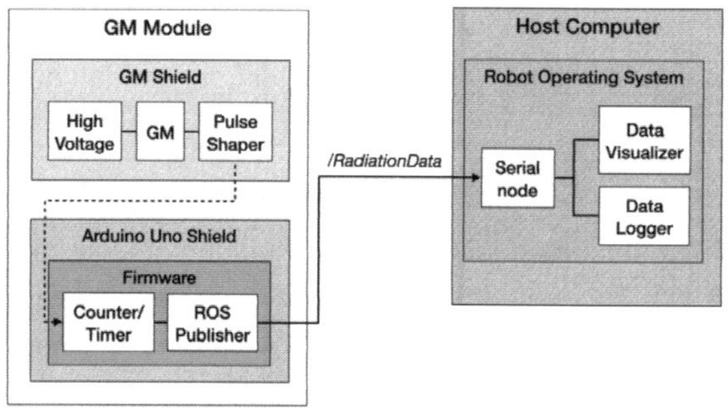

Figure 1. Block diagram of the GM module and its integration with ROS on host computer.

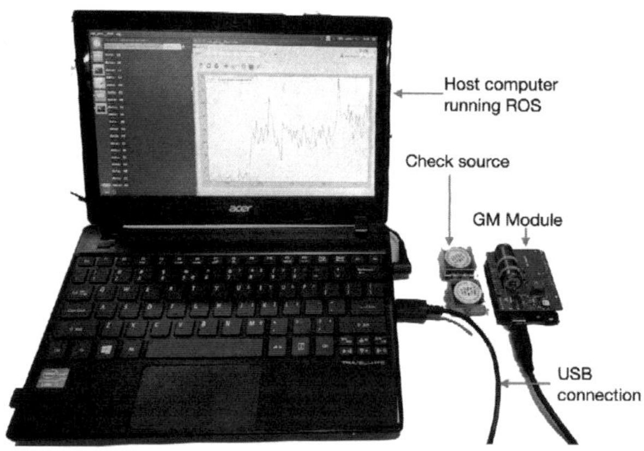

Figure 2. Physical setup and connection of the GM module with the host computer.

2.2. Design of GM module hardware

Geiger Muller (GM) detector is a gas-filled radiation detector that generates voltage or current pulses as it interacts with ionising radiation. Their terminals are connected to a high-voltage supply, creating a high-potential field within the GM chamber. Ionizing radiation that enters the GM detector will trigger gas ionisations within the chamber, leading to a Townsend avalanche and producing a voltage pulse at the GM terminals. As each incoming radiation event produces a pulse, the number of pulses per unit of time measures the intensity of the radiation field.

The GM detector circuit constructed for this system is adapted from the circuit designed in our previous work [7] . The circuit consists of the LND7121 GM tube detector, adjustable high voltage supply, and a pulse shaping circuit that converts the GM analog output to a digital TTL pulse. As shown in Figure 3, this circuit was modified as GM shield compatible to be mounted on an Arduino Uno shield.

IOP Conf. Series: Materials Science and Engineering 1308 (2024) 012007 doi:10.1088/1757-899X/1308/1/012007

Figure 3. GM module hardware.

2.3. Firmware of the GM module

The GM module firmware consist of a counter-timer program that counts the GM event detected within the counting time interval. To enable communication with ROS, the firmware utilized the ROS serial package for Arduino, *rosserial_arduino.* The rosserial ROS package uses Arduino's universal asynchronous receiver/transmitter (UART) communication and converts the GM module to a ROS node that can publish and subscribe ROS messages. The package setups and handles the ROS communication protocol as ROS topics publisher and send data from the GM module to the ROS environment of the host computer.

The firmware codes and flowchart are visualized in Figure 4. Here, a simple counter-timer is implemented to count the ionization radiation event detected by the GM detector in one second counting interval. The counter-timer output is published by the GM module as "*RadiationData*" topic. This topic can be subscribed by any ROS node of service running on the host computer.

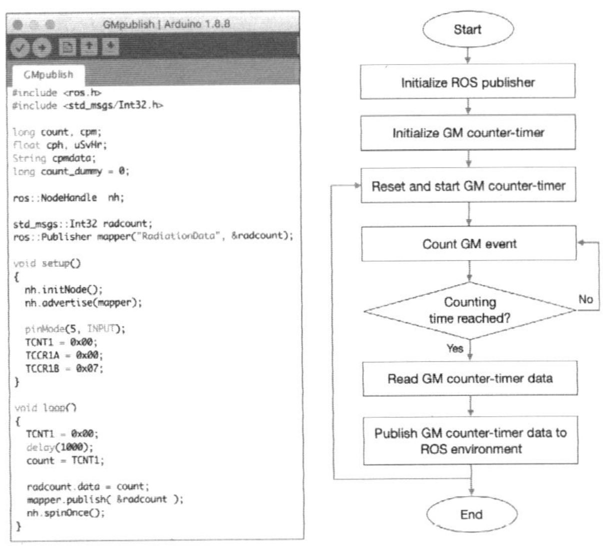

Figure 4. GM module Arduino firmware codes (left) and flowchart (right).

2.4. Setup of ROS serial interface on the host computer

On the host computer, the *rosserial_python* package was run to set up a serial node. Subsequently, the serial node could subscribe to the ROS topic */RadiationData* published by the GM module. The measured radiation data can be monitored in real-time and further manipulated or analysed from this point. The pose of the robot on 2D occupancy map is referred as the detector position. This data is acquired from the ROS *tflistener* which maintains the relationship between the *map, odom*, and *base_footprint* coordinate frames [8].

IOP Conf. Series: Materials Science and Engineering 1308 (2024) 012007 doi:10.1088/1757-899X/1308/1/012007

3. Results and discussion

3.1. GM module basic functionality test

The GM module basic functionality has been tested and verified. The high voltage circuit output was measured at 500 Voltz. Meanwhile, the output pulse of the GM detector and pulse shaper are visualized in Figure 5. The pulse shaper output is connected to the counter-timer of the GM module.

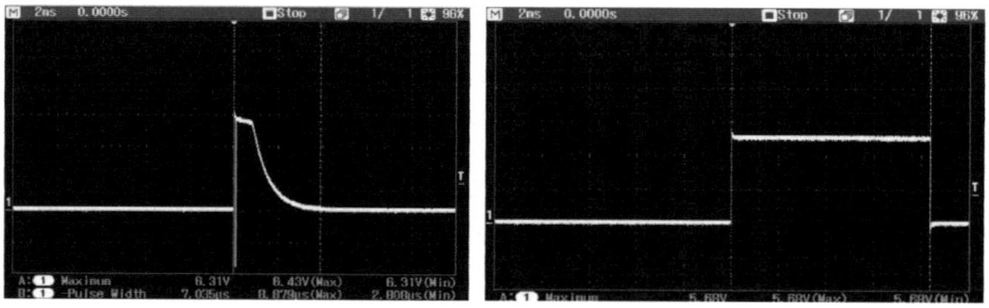

Figure 5. GM detector output and pulse shaper output

A serial node is set-up as subscriber to the ROS topic */RadiationData* published by the GM module. Data published can be viewed directly on the ROS terminal by using '*rostopic echo*' command as shown in Figure 6. Alternatively, ROS rqt viewer can be used to visualized the data in real-time. Figure 7 compares the data measured at background and with the presence of checksource on ROS rqt viewer.

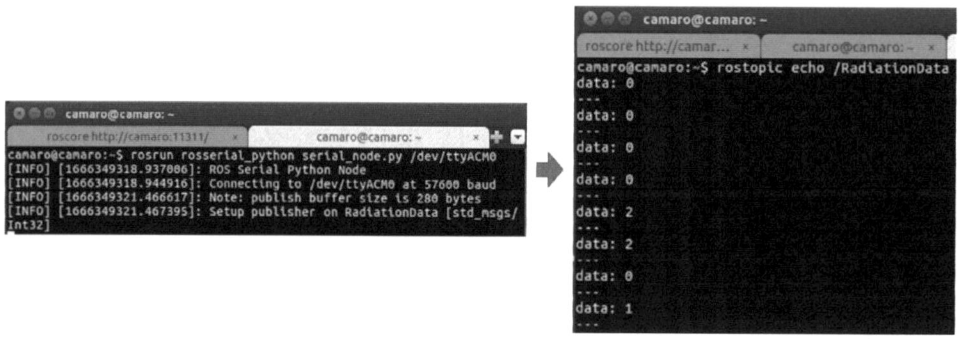

Figure 6. GM data published on */RadiationData* topic on ROS terminal (right).

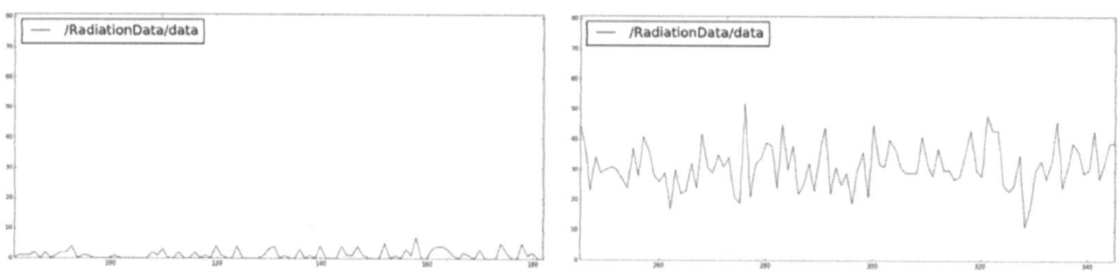

Figure 7. ROS Samples of data visualized on ROS rqt viewer for */RadiationData* published by GM module for background measurement (left) and checksource measurement (right).

3.2. *GM module data viewer and data logger*

ROS RViz tool is used as the data viewer to visualised the GM module poses, trajectories, and measured data on the ROI 2D occupancy map. Dedicated ROS nodes was programmed to create and implement RViz markers. These nodes subscribes to the */RadiationData* topic. Each published data will trigger the nodes to request the current pose from the *tflistener* to synchronise the measured radiation data and the GM detector pose. In Figure 8, the spherical markers visualised the continuous GM module i.e robot positions and trajectory. The marker colour represent the radiation intensity level, where red colour indicate high radiation intensity while green colour indicate low radiation intensity.

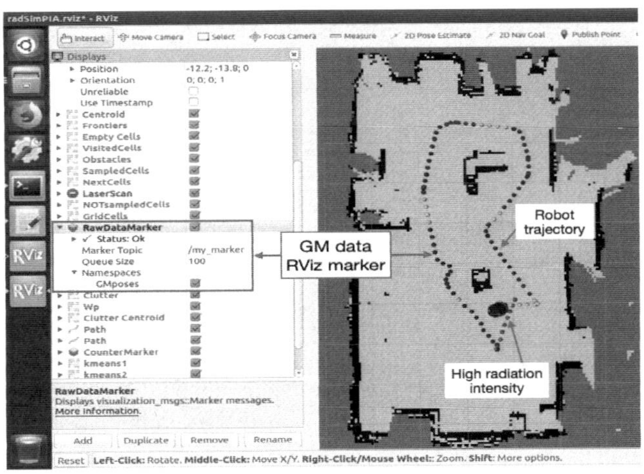

Figure 8. RViz visualisation of GM module trajectory and measured intensity.

Finally, data acquired were stored by the data logger in the log files *data_date_time.txt*. The format and sampled data of this log file is shown in Figure 9. The *data_date_time.txt* in Figure 9 (left) continuously records the position and radiation intensity data (in CPS) at 1 Hertz frequency until the node is terminated. These data are useful for inspecting or reconstructing the path of the GM module and robot during the operation.

IOP Conf. Series: Materials Science and Engineering 1308 (2024) 012007 doi:10.1088/1757-899X/1308/1/012007

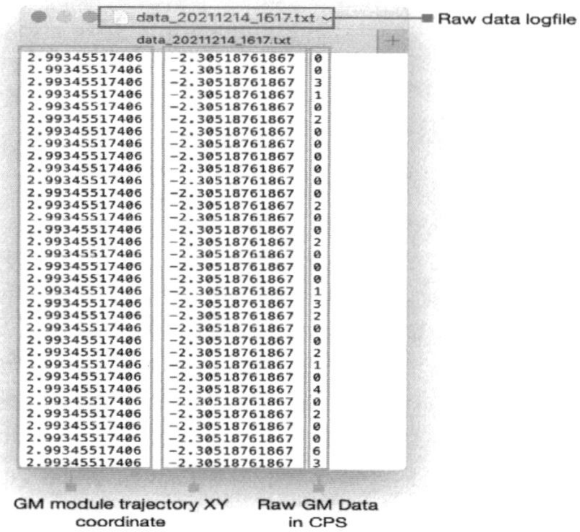

Figure 9. Log files created by the data logger.

4. Conclusion

In conclusion, this paper presented the design and integration of a radiation detector module with the Robot Operating System (ROS). The GM module incorporates a GM shield compatible with Arduino PCB. The Arduino firmware publishes data into the ROS environment, enabling the effortless visualization of radiation measurements within a 2D occupancy map. The presented knowledge in this paper could serve as the basis to enable robot to autonomously conduct radiation surveys and inspections. This advancement not only enhances the efficiency and safety of radiation workers but also contributes to safeguarding the environment against potential radiation hazards.

5. References

[1] Bird B *et al.* 2018 A robot to monitor nuclear facilities : using the continuous autonomous radiation monitoring assistance robot *IEEE Robot. Autom. Mag.* 2–10.

[2] Gabrlik P, Lazna T, Jilek T, Sladek P and Zalud L 2021 An automated heterogeneous robotic system for radiation surveys: Design and field testing *J. F. Robot.* **38**(5) 657–683.

[3] Hashem J A, Pryor M, Landsberger S, Hunter J and Janecky D R 2018 Automating high-precision X-Ray and neutron imaging applications with robotics *IEEE Trans. Autom. Sci. Eng.* **15**(2) 663–674.

[4] Marques L, Vale A and Vaz P 2021 State-of-the-art mobile radiation detection systems for different scenarios *Sensors* (*Switzerland*) **21**(4) 1–67.

[5] ROS - Robot Operating System 2021 *Open Source Robotics Foundation, Inc (Open Robotics)* https://www.ros.org/ [Accessed: 01-Jan-2023].

[6] The ROS Ecosystem 2021 *Open Source Robotics Foundation, Inc (Open Robotics)* https://www.ros.org/blog/ecosystem/.

[7] Rahman N A A *et al.* 2016 Arduino based radiation survey meter," in *AIP Conference Proceedings* **1704**.

[8] Stein M 2017 Map building and autonomous navigation using the ros navigation stack and the turtlebot - technical report for the european masters in robotics (EMARO) program at École Centrale de Nantes

Technology foresight: Capturing the benefits from nuclear science technology and innovation in Malaysian Nuclear Agency

Fairuz Suzana Mohd Chachuli[1,a)], Azlinda Aziz[1] and Faridah Mohamad Idris[1]

[1]Planning & International Relations Division, Malaysian Nuclear Agency, 43000 Kajang, Selangor

[a]fairuz@nm.gov.my

Abstract. Nuclear technology has started in Malaysia since the establishment of the Malaysian Nuclear Agency (Nuclear Malaysia) in 1972. The agency was established to promote the peaceful application of nuclear science and technology for national development. Nuclear science and technology for peaceful purposes have effectively advanced in various sectors, including industry, healthcare, food and agriculture, natural resources, energy, and nuclear safety and security through research, development, commercialization and innovation (RDCI). Scientific advancements through RDI activities will have a substantial impact on the advancement of nuclear technology. New demands placed on the public R&D infrastructure are also influencing government research priorities. Many governments believe that establishing a definite longer-term science and technology policy is essential in a time of escalating economic competitiveness. Therefore, to focus resources and achieve the greatest economic and social benefits, the most promising research fields and new technologies should be determined. Technology foresight can be used as a method to evaluate the revolutionary influence of new emerging technologies on the economy and society over the coming years, given the sustainability of RDI in nuclear technology in Malaysian Nuclear Agency. Identifying the fields of strategic research and emerging technologies with the involvement of scientists, industrialists, government officials, and others will provide the greatest economic and social advantages is the goal of technology foresight. This study proposes a conceptual model framework of the technological foresight exercise and implementation stages to be implemented in Malaysian Nuclear Agency. Through a foresight exercise, the Malaysian Nuclear Agency could close the gap between the adoption and use of nuclear technology in social and economic sectors by working with relevant industries, communities, and government agencies to match products, technologies, and services.

1. Introduction

The term of technology foresight derives from the international foresight community's shared emphasis on science and technology [1]. Several factors have contributed to the fast spread of technological foresight since the 1990s. Foresight is the process of trying to predict the long-term future of science, technology, the economy, the environment, and society. The goal is to find the areas of strategic research and new emerging technologies that are most likely to help the economy and society [2–4]. The goal of foresight is to methodically investigate potential futures. Therefore, having foresight entails adopting a proactive, aware perspective on the future and realising that decisions taken today may have a significant impact on or even create the future. The foresight initiative's impetus was driven by three main factors, which are rising governmental spending pressure, altered knowledge production processes, and increased industrial and economic competition [1].

Foresight is an umbrella term for innovative ways to plan strategies, make policies, and come up with solutions that do not predict or forecast the future but instead work with different possible futures [5]. Foresight has been described as a systematic, collaborative, future-intelligence-gathering, and medium- to long-term vision-building process that aims to help people make decisions now and work together. Foresight gives decision-makers and policy planners new ways to think about, talk about, and put into action strategic plans that are in line with the way the future is shaping up. Foresight tries to find a balance between the unsettling uncertainty and unpredictability of the future and the need for data, information, and intelligence to shape this future, without using wishful thinking, prophecies, predictions, or forecasts. Foresight helps people develop important skills like cross-impact analysis and synthesis, systems thinking, wind tunnelling and planning for deep and long-term uncertainty [5–9]. It is based on several skills, such as being aware of possible, probable, and preferred futures, scanning the horizon proactively, being able to sort, sift through, and combine open, real-time, and emerging data, and making tight feedback loops. It involves looking at possible future scenarios and paths, figuring out future risks and opportunities, and practising possible responses in a planned way.

Foresight exercises will help planning teams avoid mistakes that could cost a lot of money and get ready for different possible futures [3,6,7,10]. Foresight is widely known as a tool for making decisions [6,7]. This means that many public and private organisations can set goals and make plans that work well even when things are uncertain. Foresight exercise will objectively identify future opportunities, barriers, and risks, as well as strategically place critical thinking into long-term development. Technology foresight will also encourage participation from many different groups, which will help shape the future through policy changes.

This paper discusses the key features and characteristics of technology foresight. This led to the development of a conceptual model framework for technology foresight, which is addressed in this paper. The paper discusses the foresight initiatives in technology and innovation in Malaysia, including the proposed foresight initiatives in the Malaysian Nuclear Agency and the ten-step of technology foresight methodology. The paper also discusses the role of foresight in leveraging technology and innovation revolution that is essential for the economic and social development of the country. The paper concludes with a few key observations about technology foresight and emphasises capturing the benefits of nuclear science technology and innovation using technology foresight.

2. Key features and characteristics of technology foresight

Methodologies for foresight might be qualitative or quantitative, and they can be focused on the near or long term. The majority of foresight approaches integrate qualitative and quantitative data, thus the actual situation is not that straightforward. The periods that foresight methodologies cover and qualitative vs quantitative natures also differ. Technology foresight exercises can be carried out in a variety of ways. **Table 1** shows how a range of foresight exercises can be classified into distinct categories of foresight.

Classification of foresight might be based on the nature of the organisation involved. All organisations that support or do research should, ideally, perform foresight. Furthermore, foresight results should be neither too general nor too specific for the organisation. Foresight exercises can be characterised according to their specificity and breadth of coverage. A holistic foresight level is concerned with the complete spectrum of science and technology, whereas a macro-level foresight is typically utilised for a multi-disciplinary agency. A meso-level foresight concerning a large area or product range; and a micro-level foresight concerning a research specialism or product. To identify the convergence or fusion of formerly independent fields and the formation of new interdisciplinary areas, holistic foresight is required. However, each country must have a broad range of foresight at the macro, meso, and micro levels, as well as procedures to assure effective integration. The advantage of having a national foresight system with exercises at one level benefits from those at higher and lower levels.

Table 1. Key features and characteristics of technology foresight.

	Descriptions
Organizational characteristics	• Governmental advisory committees and major policymaking • Independent advisory councils for the public sector • Academic funding institutions • Academies and professional scientific institutions at the national and international levels • Dedicated departments and organisations • Business associations • Businesses based on science
Specificity	• Holistic • Macro-level • Meso-level • Micro-level
Roles	• Direction-setting • Choosing priorities • Predictive intelligence • Consensus invention • Advocacy • Education and Communication
Research orientation and structural characteristics	• Degree of finalisation (Curiosity-oriented vs strategic vs applied) • Complexity and stability of the disciplinary structure • External integration with wider scientific and technological systems
Inherent pressures	• Technology-push vs. demand-pull • Bottom-up vs. top-down • Third parties vs. interested parties
Time frame	• Short-term • Medium-term • Long-term
Methodology	• Formal vs. informal • Quantitative vs qualitative

The goal of technology foresight is to identify areas of research and technology that are likely to generate future socioeconomic advantages. Not only economic benefits, but also social, technology, environment, politics, and values benefits are taken into consideration while choosing prospective benefits areas. These parameters must be balanced in a way that is appropriate for the larger context in which foresight is undertaken. It is also helpful to group foresight according to its intermediary roles, which include creating priorities, generating consensus, anticipatory intelligence, advocacy, communication, and education. While priority-setting initiatives are frequently the first goal of foresight exercises, many overlook that objective. Therefore, the functions must be clearly stated at the outset to reduce the risk of an interest group takeover the exercise for their objectives. Another important lesson is that placing too much emphasis on reaching a consensus can have unintended consequences such as resource misallocation, the adoption of risk-reduction tactics, and protracted decision-making.

Technology foresight can be characterised according to its research orientation and internal disciplinary structure, whether it focuses on a mature or emerging technology. Different foresight approaches may be required for each kind. The relative importance of the various criteria will also differ. The ability to successfully balance the inherent conflict between technology-push and demand-pull forces is essential for foresight. The outcomes are typically conservative and short-term if socioeconomic demands are allowed to dominate. Two further underlying conflicts in foresight include the balance between top-down and bottom-up approaches, as well as the amount to which responsibility is assigned to an interested party or a more impartial third party. Again, balance is required for both if foresight is to be successful. Over-reliance on a top-down strategy or an independent organisation can lead to a lack of involvement and commitment from the scientific and industrial communities.

In foresight, the time horizon must be neither too long nor too short. It must be sufficiently long-term to assess developing trends and provide policymakers enough time to affect events. A foresight exercise must also not be so far into the future that researchers and others are unable to see the consequences of present-day decisions. Idealistically, the foresight time horizon should be approximately two to three times that of regular planning. Approaches to foresight can be divided into two categories based on the balance of informal and formal, as well as qualitative and quantitative methodologies. The approach chosen must be appropriate to the foresight exercise's goals and the surrounding situations. One finding here is a recent shift away from predictive and extrapolative approaches and toward the development of early-warning systems to discover new technological opportunities or demands as soon as possible.

3. Technology foresight initiatives in Malaysia

3.1. Foresight initiatives in science, technology and innovation in Malaysia

Recently, the government and industries have started to use foresight exercises to change their approach or direction in the future by using science, technology, and innovation (STI) to solve social and economic problems [11]. The combination of foresight exercises and STI would increase productivity, job creation, the ability to innovate, the pool of highly skilled workers, and, in the end, economic prosperity and the well-being of society. It would also turn the current economy, which is based on production, into a knowledge-intensive economy and push the usual limits of how things can be done. Foresight, which looks at new horizons, key drivers, and trends, as well as STI, has been recognised as an important tool for making policies and strategies related to our industrial strategies. Foresight exercises can be used as a key high-level tool for planning and strategizing so that we can stay competitive, adaptable, and quick to respond to big changes and shocks [12].

Foresight helps bring about new ideas of innovation. Foresight helps us figure out how ready we are for the future and where we can find opportunities that will lead to the solutions we need for the future. MOSTI has included STI foresight as an important strategy in the National Policy on Science, Technology, and Innovation (NPSTI) 2021-2030 [13]. This is because MOSTI knows how important foresight is, especially when it comes to building up capacity and using STI foresight. It is done to make sure that local STI foresight expertise and capabilities are expanded and strengthened. These include highlighting how STI foresight can be used to develop policies that address important issues like industrial competitiveness, food security, climate change, environmental, social, and governance (ESG) realisation, and disaster risks [14].

Sustainable Development Goals (SDG) are prioritised in Malaysian policies like the 12th Malaysia Plan, Shared Prosperity Vision 2030, NPSTI 2021–2030, and National Policy on Industry 4.0. A supportive ecosystem must be established to support the development and adoption of products and technologies that can advance the SDGs. Here, foresight is crucial in identifying the problems or gaps that prevent governments, businesses, and societies from adopting sustainable practices, as well as in creating scenarios and strategies that could be used to prevent or address those problems and gaps. Whereas traditional planning sought to prevent failure, foresight fosters agility, innovation, strategic evaluation, and the proactive shaping of the future, including supporting and creating a sustainable future for the nation. Foresight resulted in informed policy and decision-making, as well as providing confidence to the private sector and guiding them to the right investment decisions based on environmental, social, and governance factors, adopting an agile regulatory approach and stimulating RDCI activities.

The 10-10 Malaysian Science, Technology, Innovation and Economic (MySTIE) Framework combines 10 important socioeconomic factors with 10 cutting-edge scientific and technological trends that are in line with Malaysia's needs and advantages. The Emerging Science, Engineering and Technology (ESET) Study conducted by the Academy of Sciences Malaysia (ASM) was the basis for the framework, which was developed to provide S&T Foresight as part of the ASM's flagship initiative,

Envisioning Malaysia 2050 [3]. A method of foresight known as scenario planning was used to identify, characterise, and anticipate possible scenarios in Malaysia's future. By considering various alternatives for our future in 2050, policy and decision-makers, industries, communities, and STI influencers can put in place long-term planning tools for anticipating and preparing for plausible futures. The ESET study yielded 284 products, services, technologies, potential applications, and outcomes pertinent to Malaysia in 2050, 95 emerging technologies, and 21 emerging technologies with significant impact. Furthermore, by working collaboratively with the respective industries, communities, and government entities, MOSTI could bridge the gap between technology adoption and application in socio-economic sectors. Collaboration between government and industry is critical for matching products, technologies, and services from our RDCI initiatives to industrial issues, climate change, pollution, and food security.

3.2. Foresight initiatives in Malaysian Nuclear Agency

As a research institution in nuclear science and technology in Malaysia, the Malaysian Nuclear Agency (Nuklear Malaysia) plays an important role in promoting, developing and highlighting the peaceful use of nuclear technology in various fields including industry, food and agriculture, water and environment, natural resources and energy to support the sustainability of national development. Nuclear Malaysia is committed to a more competitive, productive and effective organization that performs its functions, which is firmly developed based on the foundations of nuclear science and technology. In addition, Nuclear Malaysia is also committed to empowering its core activities to align with national aspirations and international agendas including the 12th Malaysia Plan, Shared Prosperity Vision 2030, NPSTI 2021-2030 and the Sustainable Development Goals (SDG). Currently, Nuklear Malaysia is also developing the National Nuclear Technology Policy (DTNN) to plan the direction of nuclear science and technology development at the national level from 2021 to 2030. One of the initiatives proposed by stakeholders is the importance of technological foresight planning to ensure the development of the field of nuclear science and technology in Malaysia will not be left behind in the future. Activities are related to technological foresight also outlined in the Wawasan Nuklear Malaysia 2021-2030 document [15].

The combination of nuclear technology with emerging technologies such as artificial intelligence, internet-of-things, 4D printing, robotics, drones and technology will support the country's priorities in industry, medicine and health care, food and agriculture, natural resources, water and the environment in the future. In turn, this will optimize the value and impact that nuclear science and technology can offer for the country's socioeconomic development. For this reason, technological foresight is a very important element to ensure the development of new nuclear science and technology in Malaysia, especially Nuclear Malaysia so that it does not lag behind other foreign countries. Technology foresight methodology will also give focus and priority to the planning and development of new technologies related to nuclear science and technology to realize technology readiness in this country. In addition, through this technology foresight initiative, Nuklear Malaysia was able to identify the constraints it faced including issues of lack of appropriate regulatory and policy standards, human and financial resources, and research infrastructure. Accordingly, technological foresight is essential to respond to the technological challenges raised by global competition. Therefore, the proposed phases for implementing nuclear technology foresight in the Malaysian Nuclear Agency are shown in **Figure 1**.

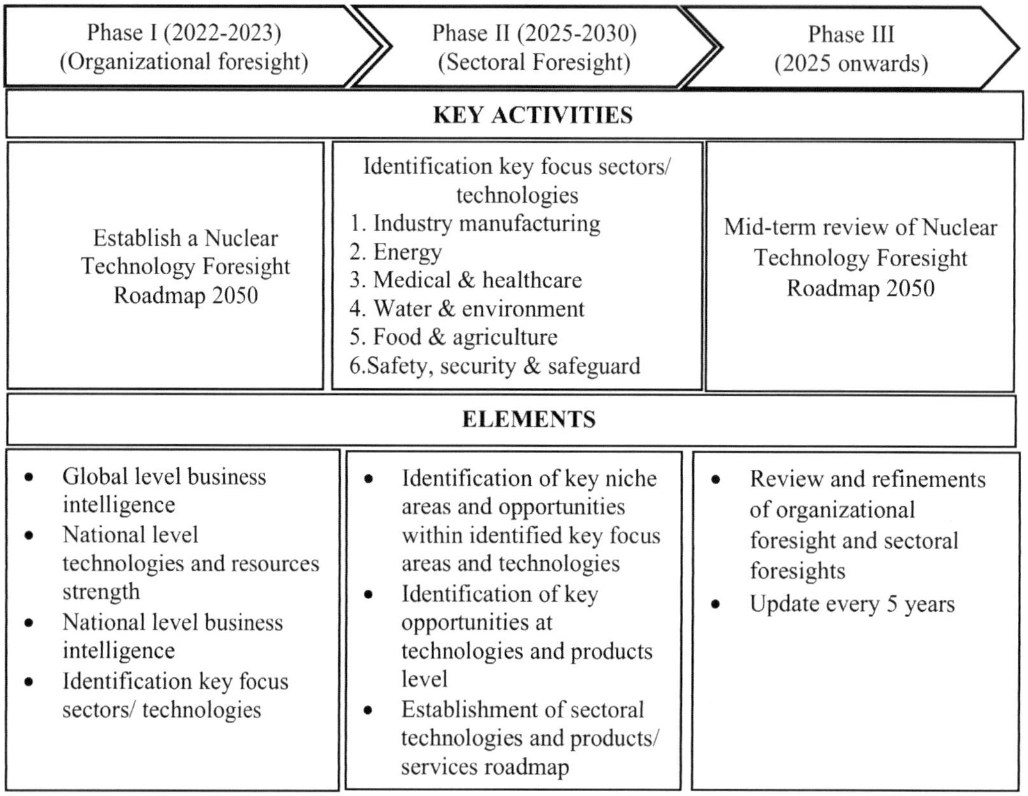

Figure 1. Proposed implementation phases on nuclear technology foresight.

3.3. Ten-step of technology foresight methodology

Future-oriented thinking mobilises change by identifying multiple possible futures to accelerate current development. Long-term transformation is embedded in future thinking, as is a clear, practical approach to cultivating a change-receptive culture. From the standpoint of development, foresight offers a variety of tools for investigating multiple futures and responding to the future's complexity with a set of rational long-term solutions [16]. The ten-step technology foresight technique depicted in **Table 2** will be used by the Malaysian Nuclear Agency in developing nuclear technology foresight exercises. The foresight exercise will be implemented using the 10-step scenario planning methodology by MyForesight, MIGHT base on the Scenario Planning Handbook [17]. The stages of technology foresight are diagnosis, prognosis, and prescription.

The first stage is the diagnosis. The diagnosis process entails assessing current conditions and trends, problems, obstacles, and factors that will either spur or hinder an organization's future transformations. Using the resources, scenarios, and approaches at our disposal, we will decide how to examine the future in the most effective way possible. There are three steps in the diagnostic process. Step 1 focuses on the project's scope, which will evolve organically with the success of the projects and produce impactful results to support future growth. In this stage, we will evaluate the strengths, weaknesses, and scope of our resources as the first step in developing our future. We will take control of our future through this process. Step 2 is to identify current challenges and trends in our immediate surroundings. The future is a small-scale version of our most extravagant desires. The future, on the other hand, provides a closer look at the environment in 10 to 20 years. Step 3 entails determining the primary change drivers. During this phase, any changes will be closely monitored, and emerging trends with the potential to replace current ones will be sought after.

The prognosis process is the second stage, which includes steps 4, 5, 6, and 7. Step 4 will highlight the main sources of uncertainty as well as several proactive measures. Significant changes that are likely to happen in the future and require prompt action are indicated by a set of signposts. Step 5 will offer a range of conceivable future scenarios that will give us previously unattainable insight into potential future opportunities. We will examine the threats and opportunities presented by these plausible scenarios. We can make it easier for people to adapt and succeed in various future scenarios by developing a specific set of strategic options that match our commitment to changing the future's scenarios. Step 6 will generate a plausible storyline that is supported by the drivers anticipated during the scenario-planning process. How the world works in the future will depend on what's happening now. The plausible scenarios are verified in Step 7, giving us well-balanced knowledge about possible futures. Each scenario's relevance to all the critical variables present will be critically assessed. By comprehending the different scenarios that the organisation may encounter, we consider our future scenario during this phase. The main themes, drivers, and actors of the scenarios are developed, and then their plausibility and logic are examined.

Steps 8, 9, and 10 make up the third stage, which is the prescription process. In the prescription phase, the organisation must determine how to translate the scenarios into strategies. Additionally, this will involve the identification of wildcards as well as future navigation and launch into the desired future. The implications are assessed in step 8 along with several strategic options for potential responses to uncertainty. To protect the future, it is important to act swiftly and rigorously in the face of shifting trends and growth drivers. Step 9 will continue to be relevant, identifying signposts and wildcards to anticipate the next major change. Tracking the progression of constantly shifting variables can be made much easier by identifying early warning signs. At this point, it is critical to identify the key indicators that will shape the future. Step 10 involves monitoring and reviewing the scenarios through a cyclical review programme to adjust the parallels with current trends. By taking this action, we will help ourselves stay focused on the future and support a prudent strategy for coordinating our future goals.

4. The role of foresight in leveraging technology and innovation revolution

Nowadays, a lot of governments believe that a clear and comprehensive technology strategy is necessary for social and economic advancement [18]. Therefore, foresight is one of the useful tools for connecting science and technology to economic and social concerns, and thus for assisting in making the appropriate decisions. The usage of foresight methodology has recently increased significantly. Government must take the lead in organising foresight throughout the full range of technology, even while the industry may oversee foresight at the meso- or micro-level [1]. To establish the context for the lower levels, a holistic foresight approach is necessary. Additionally, it provides the sole method for locating the intersection of technological streams that were previously independent. If done correctly, foresight can yield significant process benefits. It improves communication, teamwork, and shared commitment across business, academia, and government. If new technologies are to be utilised to their full potential, such links are crucial.

Initial attempts at foresight are seldom successful. A lengthy learning process is required, with considerable trial and error. There are two explanations for this. First, foresight requires weighing a huge number of frequently conflicting factors. Secondly, the absence of a coherent conceptual understanding of foresight. Looking for single-factor explanations for success and failure in foresight is futile. Successful foresight is dependent on getting most, if not all, of the factors right. This suggests a parallel with innovation theory [19]. Many lessons concerning foresight success have parallels with elements recognised in the literature as influencing success and failure in industrial innovation. Therefore, processes of innovation and foresight must be explored. In the case of foresight, these processes, such as forecast priorities, are often as significant as the outcome.

IOP Conf. Series: Materials Science and Engineering 1308 (2024) 012008 doi:10.1088/1757-899X/1308/1/012008

Table 2. The ten-step methodology of technology foresight.

	Activity
Step 1: Scoping the Project	1. Define the project's topic. 2. Set the project's boundaries and framework. 3. Determine the appropriate resources to begin a project.
Step 2: Issues and Trends Identification	List the issues and trends of the topic using the STEEPV analysis. 1. Begin with any STEEPV perspectives. 2. Consider the issues and obstacles associated with the perspective. 3. Determine trends from each perspective.
Step 3: Finding the drivers	1. Brainstorm factors that could be the driver of change based on the outcome of the list of issues and trends. 2. Identify external drivers that will shape the future landscape. 3. Identify internal drivers that will shape the future landscape.
Step 4: Identifying Key Uncertainty Drivers	1. Discuss the effect of the drivers on one another. 2. Discuss the drivers' uncertainties with each other. 3. Identify two key drivers with a high level of impact and uncertainty. 4. Brainstorm a plausible course of action for each of the ranked drivers (several options).
Step 5: The Generation of Scenarios	1. Generate characteristics of each quadrant that best characterise the future world. 2. Assign a suitable title that accurately represents the world.
Step 6: Storyline Development	1. Using the results from framing the scenario, the future world is created from the quadrant to be the Future Image 2. Describe today's situations concerning the key drivers of uncertainty to the topic 3. Brainstorm what would happen from the "Constructed Present" that leads to the realisation of the "Future Image" using a list of impactful drivers of change (How the drivers might play out in the future) 4. Discuss the interaction between every linkage and point in the Future History
Step 7: Scenarios Validation	1. Review & discuss improvements to the scenarios to make them more plausible* with adequate arguments/evidence supporting the realization of the scenarios. If not, find out what is missing. 2. Review and discuss the scenarios' storylines to make them clear and easy to understand. 3. How are the scenarios developed relevant to the institutions, sectors or countries? 4. Are the scenarios' components linked consistently? Any missing gaps?
Step 8: Assessing Implications and Outlining Potential Responses	1. Opportunities & Vulnerabilities 2. MIGHT F.I.R.S.TTM Matrix 3. SWOT analysis
Step 9: Identifying Signals	1. Identify critical Signposts / early warning signs (per scenario) that may occur in a specific scenario, such as key events, breakthroughs, and initiatives, based on the scenarios. 2. Brainstorming and/or systematic analysis of other ideas can yield complete Wild Cards. Identify the unexpected events that may occur and have an impact on your organisation in extreme circumstances. Determine a contingency or avoidance plan that can be implemented.
Step 10: Monitoring and updating the scenarios	Keeping the scenarios updated with the future environment • The scenarios are reviewed and updated every 3 years • What are the current situations that reflect the direction of the scenarios? • Does the current situation reflect the movement towards the desired direction? • What are the turns of events that could play out in 3 years? Are there any new emerging events or initiatives in the future that might trigger changes towards the scenarios' direction?

Benefits from innovation and foresight procedures must involve and gain the commitment of all those who are likely to be influenced by having a specific client with well-defined needs that are considered at an early stage [7]. Given the balance between technology-push and demand-pull viewpoints, a mix of top-down and bottom-up techniques is most likely to be successful. Foresight should not be undertaken in isolation from existing processes for planning, allocating resources, and managing research, just as R&D before industrial innovation should not be conducted in isolation from other corporate tasks [20]. Good communication is essential for innovation and foresight, both internally among all those directly involved and externally with everyone who might be impacted by the outcomes. Successful innovators keep an eye on shifting market demands and build on current advantages [6]. Analysing current strengths together with monitoring economic and social demands is another factor in a successful foresight exercise. In addition, enthusiastic persuasive people frequently play a significant role in both innovation and foresight [21].

A varied range of skills is required for the successful management of both innovation and foresight processes. A wide scientific and technical knowledge, as well as an awareness of the processes of technological development and the innovation process, are required skills for foresight [22]. Other requirements include knowledge of the sector and the needs of possible research customers, organisational and interpersonal management abilities, survey design, data-handling and analytic procedures, and a strong experience in forecasting and science-policy studies. To serve as a product champion, a manager of the innovation or foresight process, or someone else must be sufficiently senior. This will guarantee a high level of dedication to the procedure and its outcomes while allowing individuals associated with the process enough independence from conventional wisdom and immediate issues. The dissemination of the results from both processes must also be taken into consideration and means for disseminating the results to users should be included from the start. Industrial innovation is introduced through a series of iterative stages of improvement [1]. The foresight process must also be monitored, improved, and its outcomes must be validated using proper procedures.

5. Conclusions

The acceptance of foresight is constrained by organisational philosophy and the political situation. Foresight is difficult to achieve when powerful interest groups coexist and policy is set on an adversarial basis. The same holds for countries whose industrial policies prioritise the free market above long-term planning for the nation's R&D infrastructure. The independence of the scientific community is another aspect that varies by country and could limit the potential foresight. All of this is not to imply that foresight is wholly culturally specific, but rather that while creating a foresight exercise, one should consider institutional or political obstacles and opportunities. This analogy with innovation, in addition to assisting in the identification of characteristics structuring success in foresight, is a source of optimism. Given the limited success of previous attempts at foresight, some may conclude that foresight is so difficult that it is not worth pursuing at all. The example of the innovation process, which has previously perplexed many policy experts, provides hope for foresight. Foresight may be enhanced and made more effective with effort and a willingness to understand that setbacks are part of the learning process. Foresight may be enhanced and made more effective with effort and a willingness to understand that setbacks are part of the learning process. Foresight provides a method to help capture those prospective benefits and thereby shape the future by identifying the technologies likely to produce the most socioeconomic benefit.

6. References:

[1] Martin B R 1996 Technology foresight: Capturing the benefits from science-related technologies, *Res. Eval.* **6** 158–168 https://doi.org/10.1093/rev/6.2.158.
[2] WHO 2021 WHO Consultative Meeting on Science and Technology Foresight Function For Global Health 13 July 2021 Report (Geneva).
[3] ASM 2019 Science & technology foresight malaysia 2050, Kuala Lumpur, Malaysia,.
[4] TTTF, Scenario planning for transportation technology, (n.d.).
[5] Shevchenko Y, Pomogaev V, Stukach V, Nikulin D 2017 Designing and implementing future scenarios: Foresight methodology, economics, technologies, society and ecology (MPRA).

[6] UNIDO 2004 Foresight methodologies : training module 2(Czech Republic: United Nations Industrial Development Organization (UNIDO)) http://www.tc.cz/files/istec_publications/text-book-2-revised-cf.pdf.
[7] GCPSE 2018 Foresight manual empowered futures for the 2030 agenda (Singapore).
[8] Battistella C, De Toni A F 2011 A methodology of technological foresight: A proposal and field study, *Technol. Forecast. Soc. Change.* **78** 1029–1048 https://doi.org/10.1016/j.techfore.2011.01.006.
[9] ISDEFE 2021 EDA Technology foresight exercise 2021 https://eda.europa.eu/news-and-events/news/2021/09/16/eda-s-technology-foresight-exercise-delivers-first-results.
[10] CASA 2011 Strategic foresight report.
[11] Wonglimpiyarat J 2007 National foresight in science and technology strategy development, Futures. 39 () 718–728.
[12] Bereznoy A 2017 Corporate foresight in multinational business strategies (Foresight STI Gov.) https://doi.org/10.17323/2500-2597.2017.1.9.22.2017.
[13] MOSTI 2021 *National Science, Technology and Innovation Policy 2021-2030* (Putrajaya, Malaysia).
[14] MOSTI 2021 *Dasar Sains, Teknologi dan Inovasi Negara 2021-2030: Pelan Tindakan* (Putrajaya, Malaysia).
[15] Agensi Nuklear Malaysia 2021 *Wawasan Nuklear Malaysia* (Kajang, Selangor).
[16] MyForesight 2021 Transforming the world of tomorrow: reimagining the future of security and defence industries, MIGHT.
[17] MIGHT 2021 *Scenario Planning Handbook: A 10-step Practical Guide to Explore and Construct the Futures* (Cyberjaya).
[18] OECD, Strategic Foresight for Better Policies, 2019. https://www.oecd.org/strategic-foresight/ourwork/Strategic Foresight for Better Policies.pdf.
[19] Ogasawara A 2015 1st Preliminary report on the 10th science and technology foresight survey (Nistep.Go.Jp.) 1–40. http://www.nistep.go.jp/wp/wp-content/uploads/2-1_Ogasawara.pdf.
[20] Roveda C, Vecchiato R 2008 Foresight and innovation in the context of industrial clusters : The case of some Italian districts *Technol. Forecast. Soc. Change* **75** 817–833.
[21] Ruff F 2015 The advanced role of corporate foresight in innovation and strategic management — Reflections on practical experiences from the automotive industry T*echnol. Forecast. Soc. Change.* **101** 37–48.
[22] ICSTI 1999 Technology foresight Ireland: An ICSTI overview,.

Acknowledgements
The authors appreciate the support of the Planning and International Relations Division, Malaysian Nuclear Agency in this work.

Natural weathering effects on the surface morphological and physicochemical properties of radiation curable coatings in tropical climate

R Tajau[1*], MS Alias[1], MH Harun[1], NH Mudri[1], FF Hilmi[1], SM Faisal[1], RC Ismail[1], AM Sani[1], CRC Hak[2], WS Paulus[2], BC Jong[3], PWY Liew[3]

[1]Synthesis and Radiation Curing Group, Radiation Processing Technology Division, [2]Material Technology Group, Industry Technology Division, [3]Radiation Bioindustry Group, Agrotechnology and Bioscience Division, Malaysia Nuclear Agency (Nuclear Malaysia), Ministry of Science, Technology and Innovation Malaysia (MOSTI), Bangi, 43000 Kajang, Selangor.
* *rida@nm.gov.my*

Abstract. The purpose of this research is to discover the effects of natural weathering on palm oil-based and petrochemical-based film coatings. The film coatings were prepared using the radiation curing approach via cross-linking. Three palm oil-based resins with TiO_2 nanoparticle loading were developed: EPOLA, EPOLA-OPV, and POBUA. The petrochemical-based coatings were also developed to compare the performance of natural and synthetic coatings. The coatings were exposed to natural weathering for up to 60 days at varied angles of 0°, 45°, and 90°. The effect of natural weathering conditions on the surface morphological, physiochemical, and also flexural characteristics of coatings, as well as the discoloration visual inspection and the growth of mould or fungus on the tested specimen, was studied. The results showed that natural weathering exposure induced severe discoloration and deterioration of the polymer network structure of the coatings due to photo-oxidation reaction and the presence of fungi. Overall, palm oil-based coatings deteriorate slower than petroleum-based coatings, especially at a 90° angle. This study indicated that bioresources-based palm oil has a significant potential for interior wood varnish applications due to its superior qualities over synthetic-based coatings.

1. Introduction

Surface finishing materials derived from biomass resources, such as palm oil, have the potential to be one of the top choices among local industries since they have quality characteristics comparable to petrochemical-based coating products, and are even more environmentally friendly [1-4]. Along with the manufacturing of such cutting-edge materials, radiation curing technology is a green technology because it emits less or no volatile organic compounds (VOC) into the environment when compared to conventional techniques [1-7].

The integrity of the surface finishing features has emerged as a priority in the coating and paints industry and market. Natural outdoor exposure and accelerated weathering techniques are two examples of test methods developed and utilized to meet the demands of the rapid development of a novel coating and enrichment of current product qualities [8]. Coating deterioration is one of the primary research topics that have been investigated to overcome coating imperfections in various weather environments. Succeeding in fine-tuning their flaws could make the products outstanding, standing, sustaining, and globally acceptable.

Climate change is becoming a global issue that affects many industrial sectors. Recently, structural engineers have expressed concern regarding the longevity of paints and coating materials as surface protective layers in the face of global warming, which resulted in the loss of billions of dollars in assets and investments in many countries. An accelerated weathering technique estimates the performance of a coating system in a short test period; however, the actual environment following global warming should be taken into consideration when evaluating an actual atmosphere condition, particularly outdoor exposures based on country climate and location [9].

Coating deterioration can occur due to several factors, including chemical and physical processes. The chemical processes are concerned with the stability of chemical molecules and the chain network of the coating system. The physical processes are abrasions, scratches, or microorganism attacks caused by surrounding conditions and environments, such as indoor or outdoor natural exposures, atmospheric dirt, pollutants, particle contaminants, airborne particulates, microbial and insect attacks, or vandalism.

Therefore, the study aims to compare the durability studies on a series of palm oil-based and petrochemical-based UV curable coatings in Malaysia's tropical climate. The study also intends to evaluate the surface morphological and physicochemical properties of the coatings for weathered and indoor environments, as well as to analyze the coatings' fracture behavior and mould or fungus growth.

2. Materials and methods

In this study, UV-curable varnishes were used. They are palm oil-derived acrylated and urethane resins, specifically EPOLA, EPOLA-OPV, and POBUA, whereas the petrochemical-derived resins are urethane acrylates produced at the Synthesis and Radiation Curing Laboratory, Radiation Processing Technology Division (Nuclear Malaysia) (Table 1) [1]. The three types of coating substrates are glass, rubber wood, and cemboard. The coatings were applied by brushing with thicknesses ranging from 50 to 150 μm and exposed to UV radiation (IST UV machine) at a conveyor speed of 10m/min and a current of 7.5 amperes. The specimens are placed outside and oriented at different angles, namely 0°, 45°, and 90°, in accordance with the standard method for natural weathering of materials (ASTM G7/G7M-21) (Fig. 1). The study site is in MINT-Tech Park, Nuclear Malaysia, Dengkil, Selangor, with a latitude of 2.9001094 (north) and a longitude of 101.7595219 (east), indicating a tropical climate environment. Visual inspections were performed after 30 and 60 days. An FTIR spectrophotometer (Bruker's Tensor II) is used to analyze chemical functions. The optical contact angle (Biolin Scientific, TL 100) was utilized to determine the hydrophobicity of the coating. The chemical resistance test followed ASTM D 5402-93 (1999) and used the standard method of solvent resistance of coatings utilizing the solvent rub technique. The FESEM images were collected using a Zeiss microscope (model Philips XL40). The glossiness of the coated films was measured using an ASTM D-523 Micro-TRI-Gloss meter. Antibacterial activity on plastic and non-porous surfaces is measured using a modified method based on ISO 22196:2011, which has been used to carry out antimicrobial assessments. The palm oil-based coating films were applied to agar plates that had previously been seeded with the following bacteria: *Escherichia coli and Bacillus subtilis*. A swab and dip test were also performed on the surface of coatings samples that had been stored at room temperature for a year and a half.

Table 1. Coating types and sample identification.

Sample No.	Palm oil-based coatings	Sample No.	Petrochemical-based coatings
S1	EPOLA	S9	Hard coating
S2	EPOLA-OPV	S10	Hard coating-TiO_2 NPs
S3	EPOLA-OPV (Dissolution)	S11	Hydrophobic coating
S4	POBUA	S12	Hydrophobic coating-TiO_2 NPs
S5	EPOLA-TiO_2 NPs		
S6	EPOLA-OPV-TiO_2 NPs		
S7	EPOLA-OPV-TiO_2 NPs (Dissolution)		
S8	POBUA-TiO_2 NPs		

Figure 1. Natural weathering testing at the MINT-Tech Park research site in Dengkil, Selangor.

3. Results and discussion

3.1. Hydrophobicity

The contact angle of water droplets on the palm oil-based coating surface increases with the presence of TiO_2 nanoparticles (NPs) and these surfaces are defined as hydrophobic with contact angles larger than 90° (Fig. 2b). Similarly, the findings of the Ismail et al. investigation demonstrate that the presence of silica particles increases the hydrophobicity of the coating, with a water contact angle attained over 90° [4]. In the addition of TiO_2 NPs, the thin layer of these cross-linked coatings may repel water (Fig. 2c). This action is referred to as dewetting since the water pushes away from the surface, leaving it completely dry.

The hydrophobic character of the coatings with TiO_2 NPs addition could minimize water molecule penetration and absorption into the coated substrates. These coatings attain their high degree of water repellence by a chemical interaction of highly nonpolar molecules in the coating system, which align and create the coating's surface. As a result, TiO_2 coatings are an ideal choice for the most demanding applications and environments.

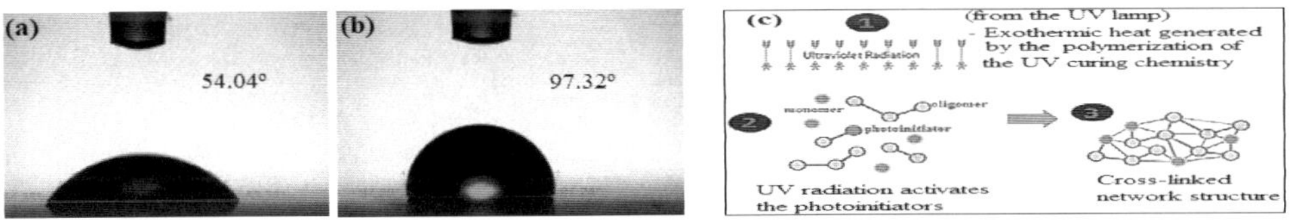

Figure 2. Water contact angle of coatings in swollen states: (a) S2, (b) S6 and (c) cross-linked network in polymer coatings.

3.2. Natural weathering effects

Surface deterioration caused by natural weathering on the coatings reduced the surface morphological and physiochemical qualities, as evidenced by visual observation and FTIR spectroscopy. The coating deterioration is a combination of physical and chemical processes.

3.2.1. Surface morphological and physical flexural analysis. Significant reductions in surface morphological features of the uncoated wood substrate where mould formed were found, indicating that the effects of weathering conditions are more visible on uncoated surfaces than on coated surfaces (Fig. 3-5). The alternating dry and wet conditions may trigger the development of osmotic molecules in the coating, destroying coating systems and boosting substrate damage and mould growth [8]. The coatings were significantly damaged as a result of the weathering impact caused by film coating degradation after 30 and 60 days of exposure to natural weathering (Fig. 3-5). As a result, uncoated surfaces could not withstand microbial growth and dirt for more than 30 days (Table 2).

Figure 6b illustrates the TiO_2 coating non-uniformities presented by the FESEM image. The TiO_2 NPs are randomly distributed on the coated surfaces, with uneven particle size and spreading as shown on the film layer. The presence of these non-uniformities may have an impact on the protective layer's functionality.

Table 3 examines the physical impacts of 30 days of exposure to environmental weather on cemboard, glass, and wood surfaces. The coating material's adhesion to the glass surface deteriorates after 7 days of exposure, whereas the coating material's adhesion to the wood and cemboard surfaces remains good for 30 days (Table 3). When compared to a more exposed surface without a coating, the advantage of the hydrophobic nature of the coating's resistance can be achieved up to the 30th day before the coating layer begins to deteriorate.

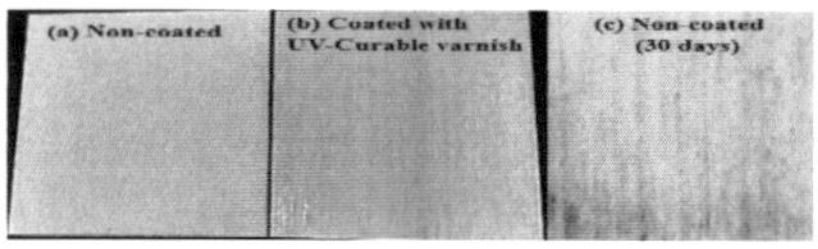

Figure 3. Comparison of wood surfaces: (a) Original wood surface without protective coating; (b) Coated wood surface and (c) Effect of 30 days of weather exposure on the uncoated wood surface without a protective coating.

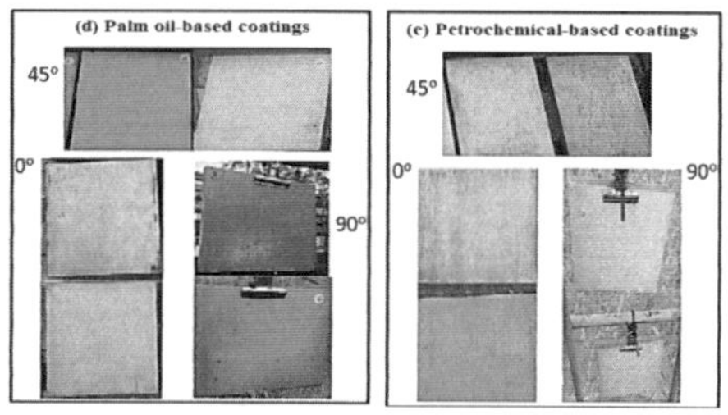

Figure 4. The effect of sample exposure on the weather after 30 days at different exposure angles of 0°, 45° and 90°.

Figure 5. The physical condition of the sample after 60 days at various angles of 0°, 45° and 90°.

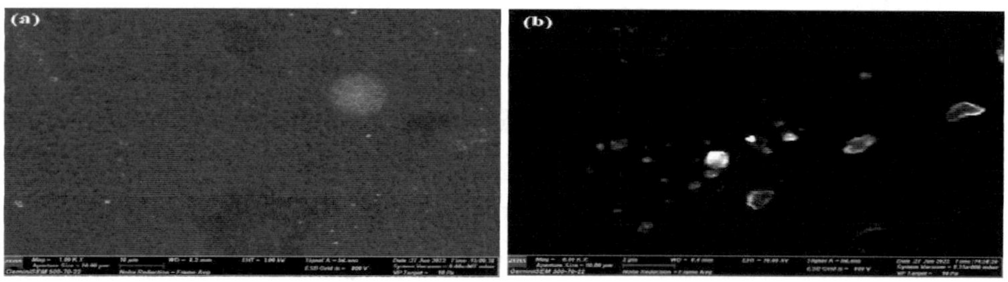

Figure 6. FESEM images of palm-oil based coatings: (a) S1-neat coating and (b) S5-with TiO_2 NPs.

Table 2. Visual inspection of the microbe growth

Sample	S1			S2			S3			S4			S5			S6			S7			S8			S9			S10			S11			S12		
DAY 1	W	C	G	W	C	G	W	C	G	W	C	G	W	C	G	W	C	G	W	C	G	W	C	G	W	C	G	W	C	G	W	C	G	W	C	G
0°	N	N	N	N	N	N	N	N	N	N	N	N	N	N	N	N	N	N	N	N	N	N	N	N	N	N	N	N	N	N	N	N	N	N	N	N
45°	N	N	N	N	N	N	N	N	N	N	N	N	N	N	N	N	N	N	N	N	N	N	N	N	N	N	N	N	N	N	N	N	N	N	N	N
90°	N	N	N	N	N	N	N	N	N	N	N	N	N	N	N	N	N	N	N	N	N	N	N	N	N	N	N	N	N	N	N	N	N	N	N	N

Sample	S1			S2			S3			S4			S5			S6			S7			S8			S9			S10			S11			S12		
DAY 7	W	C	G	W	C	G	W	C	G	W	C	G	W	C	G	W	C	G	W	C	G	W	C	G	W	C	G	W	C	G	W	C	G	W	C	G
0°	N	N	N	N	N	N	N	N	N	N	N	N	N	N	N	N	N	N	N	N	N	N	N	N	N	N	N	N	N	N	N	N	N	N	N	N
45°	N	N	N	N	N	N	N	N	N	N	N	N	N	N	N	N	N	N	N	N	N	N	N	N	N	N	N	N	N	N	N	N	N	N	N	N
90°	N	N	N	N	N	N	N	N	N	N	N	N	N	N	N	N	N	N	N	N	N	N	N	N	N	N	N	N	N	N	N	N	N	N	N	N

Sample	S1			S2			S3			S4			S5			S6			S7			S8			S9			S10			S11			S12		
DAY 30	W	C	G	W	C	G	W	C	G	W	C	G	W	C	G	W	C	G	W	C	G	W	C	G	W	C	G	W	C	G	W	C	G	W	C	G
0°	Y	Y	-	N	S	-	S	N	-	S	Y	-	Y	Y	-	S	N	-	Y	N		Y	Y	-	-	-	-	-	-	-	Y	Y	-	Y	Y	-
45°	Y	S	-	S	S	-	Y	Y	-	Y	Y	-	Y	Y	-	N	N	-	Y	Y		Y	Y	-	-	-	-	-	-	-	Y	Y	-	Y	Y	-
90°	S	S	-	N	S	-	S	Y	-	S	Y	-	S	N	-	S	N	-	S	Y		N	N	-	-	-	-	-	-	-	S	Y	-	S	Y	-

W Wood; C Cemboard; G Glass.

Indication of mould growth; Y Yes; N No; S Slightly.

Table 3. Visual inspection of adhesion.

Sample	S1			S2			S3			S4			S5			S6			S7			S8			S9			S10			S11			S12		
DAY 1	W	C	G	W	C	G	W	C	G	W	C	G	W	C	G	W	C	G	W	C	G	W	C	G	W	C	G	W	C	G	W	C	G	W	C	G
0°	√	√	√	√	√	√	√	√	√	√	√	√	√	√	√	√	√	√	√	√	√	√	√	√	√	√	√	√	√	√	√	√	√	√	√	√
45°	√	√	√	√	√	√	√	√	√	√	√	√	√	√	√	√	√	√	√	√	√	√	√	√	√	√	√	√	√	√	√	√	√	√	√	√
90°	√	√	√	√	√	√	√	√	√	√	√	√	√	√	√	√	√	√	√	√	√	√	√	√	√	√	√	√	√	√	√	√	√	√	√	√

Sample	S1			S2			S3			S4			S5			S6			S7			S8			S9			S10			S11			S12		
DAY 7	W	C	G	W	C	G	W	C	G	W	C	G	W	C	G	W	C	G	W	C	G	W	C	G	W	C	G	W	C	G	W	C	G	W	C	G
0°	√	√	F	√	√	F	√	√	F	√	√	F	√	√	F	√	√	F	√	√	F	√	√	F	√	√	√	√	√	√	√	√	F	√	√	F
45°	√	√	F	√	√	F	√	√	F	√	√	√	√	√	F	√	√	F	√	√	F	√	√	F	√	√	F	√	√	F	√	√	F	√	√	F
90°	√	√	√	√	√	F	√	√	F	√	√	√	√	√	F	√	√	F	√	√	F	√	√	√	F	F	√	F	F	F	√	√	F	√	√	√

Sample	S1			S2			S3			S4			S5			S6			S7			S8			S9			S10			S11			S12		
DAY 30	W	C	G	W	C	G	W	C	G	W	C	G	W	C	G	W	C	G	W	C	G	W	C	G	W	C	G	W	C	G	W	C	G	W	C	G
0°	√	√	F	√	√	F	√	√	F	√	√	F	√	√	F	√	√	F	√	√	F	√	√	F	√	√	F	√	√	F	√	√	F	√	√	F
45°	√	√	F	√	√	F	√	√	F	√	√	F	√	√	F	√	√	F	√	√	F	√	√	F	√	√	F	√	√	F	√	√	F	√	√	F
90°	√	√	F	√	√	F	√	√	F	√	√	F	√	√	F	√	√	F	√	√	F	√	√	F	√	√	F	√	√	F	√	√	F	√	√	F

W Wood; C Cemboard; G Glass.

Indication of cracking and adhesion: √ Good; F Failure.

3.2.2. FTIR. The FTIR test findings demonstrated that after 30 days of natural weathering exposure, the palm oil / bio-based coatings (S2) (Fig.7a) and petrochemical-based coatings (S11) (Fig.7b) were slowly degraded. Significant reductions in chemical functional group properties of coatings where the polymer structure cleaved were observed for the weathered specimen, which reduced drastically at 0° and 45° for the bio-based/palm oil-based coating, indicating that the effects of weathering conditions are more pronounced at 0° and 45° than at 90° (Fig.7a).

The chemical functional groups of the coatings' carbonyl, ester, aromatic, carbon double bonds, alkane, and hydroxyl peaks reduced upon exposure. The coating deterioration was induced by external environmental factors such as humidity, heat, and UV radiation [8,10]. The degradation by hydrolysis promotes water molecules to absorb into the coating system, attack the functional groups of the polymer chains, and cleave the chemical structure network (Fig.8a), resulting in a primary deterioration of the protective coating layer that allowed interfaces between water molecules and the substrate surface [8,10].

Long-term exposure of coating systems to sunlight or UV rays also contributed to the photo-oxidation phenomenon on the coating system, which built the deterioration process (Fig.8b). The surface morphology of these coating systems deteriorated after 30 and 60 days, causing mould formation (Fig. 3-5) [10].

The palm oil-based coating, on the other hand, decays more slowly than petrochemical-based coatings after 30 days of exposure to natural weathering (Fig.8c). The amine group -NH at 1526 cm^{-1} and -C-CO-C- at 1071 cm^{-1} was degraded for the model compound of urethane and ester carbonyl, with the peak intensity decreasing after up to 30 days of exposure, as reported by Yang et al., who found that urea and urethane group concentrations increased and caused blisters on the coating after exposure [8].

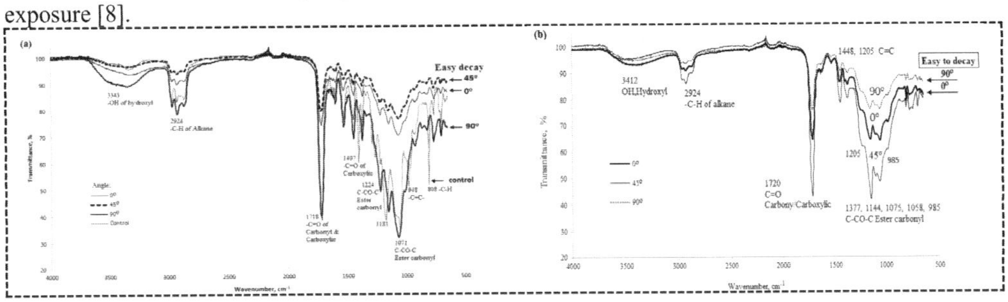

Figure 7. The FTIR test results of (a) the palm oil / bio-based coatings (S2) and (b) the petrochemical-based coatings (S11) after 30 days of natural weathering exposure.

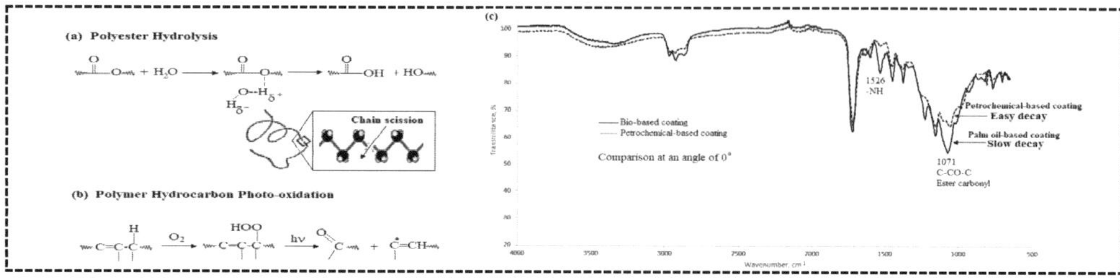

Figure 8. Coating degradation in polymer structure: (a) hydrolysis, (b) photo-oxidation and (c) comparison of the FTIR spectra of the bio-based/ palm oil-based and the petrochemical-based coatings after being exposed to natural weathering for 30 days.

3.3. Indoor temperature effects

The ester carbonyl functional group at 1071 cm^{-1} caused the coating system to deteriorate slowly (Fig. 9). Controlling the environment, such as room temperature and avoiding direct sunlight and water exposure, makes the coating with a cross-linked structure exceptionally durable after one and a half years [9]. Furthermore, a rub test performed using chemical solvents such as MEK, ethanol, and acetone resulted in the initial removal of the coating (Table 4). On the other hand, the coating glossiness increases as shown on the wood surface after rubbing with water, which complements the water molecules' hydrophobicity (Fig. 2) [4,9]. However, after one and a half years, the coating glossiness measured decreased.

Table 4. The effect of a rub test on the glossiness of a coated-wood surface.

Sample	Rub Test (%) (before and after solvent and water rub test (50 rubs) on wood surface at angle of 20°)							
	Methyl Ether Ketone (MEK)		Etanol		Acetone		Water	
	Before	After	Before	After	Before	After	Before	After
S2	44.1	38.6	27.4	31.9	33	25	39.7	53.2
S2 (after 11/2 years)	10	9.8	16.5	14	18.2	13.2	22.6	25.2

3.4. Antimicrobial assessments

The disc method demonstrated no effect of microbial inhibition on the top and bottom of the film after the third and seventh days (Fig. 10a). There was also no microbial growth on the surface of 1-and-a-half-year-old coatings, according to swab and dip tests (Fig.10b). The UV-curing cross-linking significantly increased the surface properties of the coating, resulting in exceptionally effective antimicrobial properties that are reasonably long-lasting [6-7]. Overall, this coating provides excellent resistance to moisture and atmospheric dirt for interior applications [9].

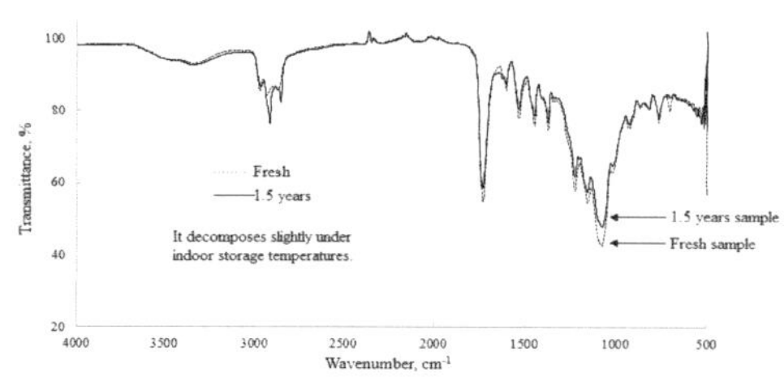

Figure 9. The effect of deterioration of the chemical structure of the coating material at room temperature for more than 1 year (S2).

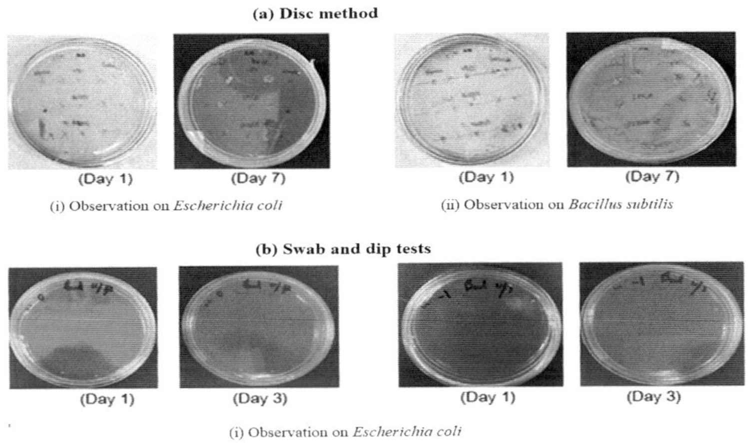

Figure 10. Antimicrobial assessments of the palm oil-based coating films with tested bacteria of *Escherichia coli* and *Bacillus subtilis*, (a) Disc method and (b) Swab and dip tests.

4. Conclusion

Palm oil-based coatings have the potential to be used in comparable applications such as wood-based material due to their comparable qualities to synthetic petrochemical-based coatings. It was suggested that a suitable primer and base layer treatment be applied to the substrate before top coating with palm oil-based varnishes to improve its durability for widespread application in various industries, particularly surface finishing for wooden surfaces, including furniture. It is waterproof, UV resistant, antimicrobial, and enhances the aesthetics of wood in interior applications.

5. References

[1] Tajau R, Rohani R, Alias MS, Mudri N H, Abdul Halim K A, Harun M H, Mat Isa N, Che Ismail R, Muhammad Faisal S, Talib M, Mohamed Zin M R , Izni Yusoff I, Khairul Zaman N and Asyila Ilias I 2021 Emergence of polymeric material utilising sustainable radiation curable palm oil-based products for advanced technology applications *Polymers* **13** 1865.

[2] Alias M S, Othman N, Kamarudin S, Harun M H, Mohamed M, Mohamad S F, Saidin N, Tajau R, and Abdul Halim K A 2022 Green corrosion coating using palm oil based urethane acrylate (POBUA) on mild steels in 3.5 NaCl solution *Materials Today: Proc.* **66**.

[3] Abdul Halim K A, Malek Faisal A A, Che Ismail R, Mohd Sani A M, Mudri N H and Mat Isa N 2023 Effects of NCO/OH ratio on the properties of radiation curable waterborne polyurethane acrylate *e-Jurnal Sains Nuklear Malaysia* **35(1)** 67 – 76.

[4] Ismail N B, Jamaluddin A N, Norizham N A, Mohd Nor N H, Othman N, Alias M S, Abdul Halim K A, Mohamed M and Harun M H 2021 Characterization of hydrophobic UV-curable acrylated coating from palm oil besed urethane acrylate (POBUA) for wood coating application *J. of Nucl. and Related Technol.* **(18)2** 1-7.

[5] Mudri N H, Abdullah L C, Aung M M, Biak D R A and Tajau R 2021 Structural and rheological properties of nonedible vegetable oil-based resin *Polymers (Basel)* **13(15)** 2490.

[6] Shen Z, Wu Y, Qiu S, Deng H, Hou R and Zhu Y 2020 UV-thermal dual-cured polymers with degradable and anti-bacterial function *Progress in Organic Coatings* **148** 105783.

[7] Noè C, Tonda-Turo C, Carmagnola I, Hakkarainen M, Sangermano M 2021 UV-cured biodegradable Methacrylated starch-based coatings *Coatings* **11(2)** 127.

IOP Conf. Series: Materials Science and Engineering 1308 (2024) 012009 doi:10.1088/1757-899X/1308/1/012009

[8] Yang X F, Vang C, Tallman D E, Bierwagen G P, Croll S G, Rohlik S 2001 Weathering degradation of a polyurethane coating *Polym. Degrad. and Stability* **74(2)** 341-351.
[9] Nguyen T V, Do T V, Ha M H, Le H K, Le T T, Nguyen T N L, Dam X T, Lu L T, Tran D L, Vu Q T, Dinh D A, Dang T C and Nguyen-Tri P 2020 Crosslinking process, mechanical and antibacterial properties of UV-curable acrylate/Fe3O4-Ag nanocomposite coating *Progress in Organic Coatings* **139** 105325.
[10] Hashim U R, Jumahat A, Jawaid M, Dungani R. and Alamery S 2020 Effects of accelerated weathering on degradation behavior of basalt fiber reinforced polymer nanocomposites *Polymers* **12(11)** 2621.

Acknowledgements
The Authors would like to thank Radiation Processing Technology Division, Industry Technology Division and Agrotechnology and Bioscience Division, Malaysia Nuclear Agency, and Ministry of Science, Technology and Innovation Malaysia for the research support. The authors would like to extend their gratitude to Nuclear Malaysia for the funding of this research through Researchers Project numbers (RIMC: NM-R&D-20-52 and NM-R&D-22-96).

IOP Conf. Series: Materials Science and Engineering 1308 (2024) 012010 doi:10.1088/1757-899X/1308/1/012010

Attenuation coefficient for Agarwood and shade tree sections at 59.6keV and 356keV

Susan Maria Sipaun[1,a)], **Mohd Fakhrul Hidayat Bin Anuar**[2] **and Hearie Bin Hassan**[1]

[1]Agensi Nuklear Malaysia, Bahagian Teknologi Industri, 43000 Bangi, Selangor
[2]Universiti Malaysia Sabah, Fakulti Sains dan Sumber Alam, 88400 Kota Kinabalu, Sabah

[a]susan@nm.gov.my

Abstract. Gamma ray measurements were carried out on wood sections of agarwood and shade trees to obtain its mass attenuation coefficients. Diameter of the blocks of agarwood (Aquilaria Malaccensis) is 17.0cm and shade tree (Khaya Senegalensis) sections are 24.0cm (branch) and 37.2cm (trunk). The radiation transmission measurement setup used a 0.5" x 0.5" thallium-activated sodium iodide detector, scaler ratemeter Ludlum model 2200 and collimated radioactive sources 3.5GBq Am-241 and 0.05GBq Ba-133 for the attenuation measurements. The dry density of agarwood trunk was 0.40±0.01 gcm^{-3}. For the shade tree, the dry densities were 0.52±0.02 gcm^{-3} (branch) and 0.49±0.01 gcm^{-3} (trunk). At 59.6keV, the mass attenuation coefficient for the agarwood at diameter is 0.114±0.007 cm^2g^{-1}, and for shade tree(branch) is 0.136±0.004 cm^2g^{-1} and 0.120±0.004 cm^2g^{-1} shade tree (trunk). At 356keV, the mass attenuation coefficient for the agarwood at diameter is 0.081±0.003 cm^2g^{-1}, and for shade tree (branch) is 0.086±0.004 cm^2g^{-1} and 0.068±0.028 cm^2g^{-1} shade tree (trunk).

1. Introduction

Previous studies reported various ways to determine wood attenuation coefficient and density using water immersion/density metric method, photon and x-ray measurements, ultrasonic goniometry as well as MCNP calculations [1-10]. Density and strength of wood indicate the quality of wood [11]. It has been established that these efforts provide a way to identify particular types of wood, including Rhizophora spp, Agarwood, soft and hard wood, Pinus Radiata and Tectona Grandis through direct measurements or indirectly [1-6, 9]. The mass attenuation coefficient is a measure of the probability of the interaction that occurs between incident photons and the matter of the unit mass per unit area.The factors that affect attenuation are related to the incident photon beam and the properties of the material through which the radiation traverses. These factors include the incident beam energy, the thickness, atomic number and the density of the material. The density of wood, exclusive of water, varies greatly both within and between species. In this present study, we describe the measurements of mass attenuation coefficients for local shade tree and agarwood tree sections both in dry and wet conditions. Dry densities are recorded as 0.40±0.01, 0.49±0.01 and 0.52±0.02 gcm^{-3} for agarwood, shade tree(branch) and shade tree(trunk) respectively. Both Am-241 and Ba-133 are suitable gamma ray radioactive sources due to their excellent penetrability and sensitivity through low-density materials. This study provides the density data to use in gamma densitometry whereby it may be used to reasonably estimate locations of gaharu resin within a tree. Classification of agarwood (gaharu) by resin content has been described by MA Nor Azah *et.al* [12]. Agarwood is valued for its distinctive fragrance and highly valuable fragrant heartwood [12-13]. Due to its pharmalogical properties, the species of Aquilaria spp also provides health applications from the use of its leaves, stem, bark and wood. This includes having antioxidant, anti-bacterial and antimicrobial effects [14].

2. Material and methods

Woods sections were first measured and weighed to obtain wood density for both shade tree and agarwood samples. This was done for both dry sample and wet sample. The shade tree sample came from a species of Khaya Senegalensis, a common shade tree planted for landscaping in Malaysia [15]. Measurements for this wood was performed within six months from the day the tree was cut down, however the agarwood sample is at least 10 years old. The shade tree density is approximately 20-30% higher than agarwood density for this case. Table 1 shows the physical properties of the wood samples.

Table 1. Physical properties of wood samples.

Wood identifier	Density-dry (gcm^{-3})	Density-wet (gcm^{-3})	Diameter (cm)	Weight dry (gram)	Weight wet (gram)	Sample
A: Khaya (branch)	0.52±0.02	0.57±0.02	24.0 ±0.1	1238.5 ± 0.5	1667.0 ± 0.5	
B: Agarwood	0.40±0.01	0.70±0.02	17.0 ±0.1	1565.5 ± 0.5	2827.0 ± 0.5	
C: Khaya (trunk)	0.49±0.01	0.63±0.01	37.2 ±0.1	7539.0 ± 0.5	10920.5±0.5	

For sample A, the wood was intact with no obvious signs of rotting or damage to the wood. Data was first collected in this intact condition. Measurements were taken for two (2) conditions; dry and wet. Wet condition is achieved by fully immersing the wood in plain water for 24 hours. Next, we drilled holes into the wood to create voids within the section. The holes were 2mm, 5mm and 10mm in diameter. Measurements were repeated in the same experimental setup to detect small voids within the wood. The detection of the smallest voids indicates the sensitivity of the gamma ray transmission technique in detecting density differences within the solid wood. This step is useful in simulating detection of smallest resin formation within an Agarwood tree.

For sample B, the Agarwood was acquired from an Agarwood plantation. The resin has been taken out for processing leaving a hole within the wood section. This sample has been kept for more than 10 years in room temperature storage unit and is dry. Measurements were taken for two (2) conditions; dry and wet. Wet condition is achieved by fully immersing the wood in plain water for 24 hours.

For sample C, the wood has some rotting in the middle area. This section was cut from the bottom section of the tree trunk. Masking tape was used to keep the internals in its original position and for marking positions. Measurements were taken for two (2) conditions; dry and wet. Wet condition is achieved by fully immersing the wood in plain water for 24 hours.

Background count measurements were taken before and after experiments. These readings were taken to monitor experimental conditions (electronics and environment) were maintained throughout the experiment. Counting time on the scaler/ratemeter was set at 6 seconds for all data collection. The dial setting (DS)/high voltage (HV) setting is adjusted accordingly for each gamma source and for each cable-scaler/ratemeter setup combination. The sodium iodide detector was placed in a lead holder that has a 5mm dia. collimator. The radioactive source was also placed in a lead holder with a 15mm x 5mm collimator. Collimation prevents unnecessary exposure of sample and equipment that are outside the area of interest. The source-detector were aligned and directly facing each other so that the counting system measures the decrease in intensity along a linear path. Measurements were taken at every 1cm interval so as to obtain a series of intensities across the wood section. Three readings were taken for each point (linear path) and corrected for background.

Two radioisotopes were used in this work; Am-241 (59.5409 keV) and Ba-133 (356keV). We studied the gamma ray attenuation of wood using different radioisotope energies and activities along its cross section. The detection sensitivity for voids (holes) within the wood section is examined. Each sample was exposed to both Am-241 and Ba-133 separately and attenuation data were collected and analysed. The averaged background and air counts are listed in Table 2.

Table 2. Averaged counts for samples A, B and C in their respective setups.

	Air Count (counts/6s)			Background (counts/6s)		
Radioisotope	**A**	**B**	**C**	**A**	**B**	**C**
Am-241	48684	50047	28786	17	19	20
Ba-133	4214	4199	2536	16	19	21

The mass attenuation coefficient, μ_m is given by the division of the linear attenuation coefficient, μ and the density, ρ.

$$\mu_m = \frac{\mu}{\rho} = x^{-1}.\ln\left(\frac{I_o}{I}\right) \quad (1)$$

The counts represent intensity of transmitted beam, I and the mass attenuation coefficient, μ_m in equation (1) was calculated and tabulated. These measurements were taken for solid wood at thickness range A: 3.6-24.0cm, B: 4.2-16.6cm, C:.4.1-35.8cm. Mass attenuation coefficient, μ_m is calculated and tabulated. Special attention is given to mass attenuation coefficient values at diameter to compare material discrimination ability between Ba-133 (higher energy) and Am-241 (lower energy). Figure 1 shows the experimental setup for the measurements.

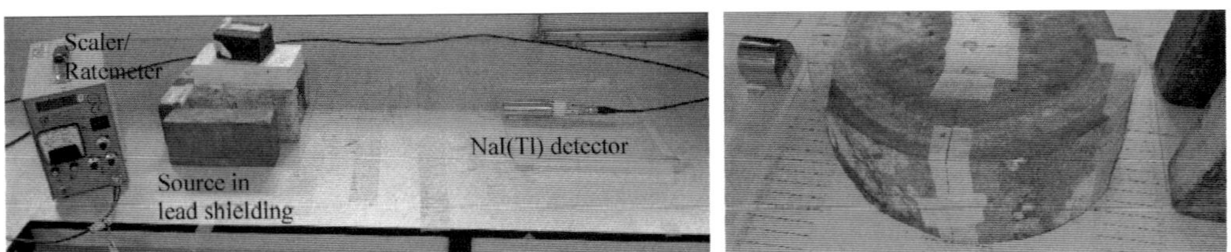

Figure 1: Source-detector arrangement for measurement of the wood sections. Taking background and air count measurement (left). Taking sample count measurement (right)

3. Results and discussion

The count measurement results are presented in figures 2-5 below. Figures 2 and 3 show the results of density measurement on sample A using Ba-133 and Am-241 respectively. Each figure gives the experimental results for measurement conditions that is dry without void, dry with void and wet with void. The baseline data is taken as the result obtained for dry without void; wood section is intact in original shape.

Sample B (17.0 cm dia.) is taken from a section of an Aquilaria tree. This sample is dry, not freshly cut from the tree and has been kept in the store for many years. The gaharu resin has been taken out and left a void. Its measurement results for both Ba-133 and Am-241 is shown in figure 4. There are two sample conditions, dry and wet. For both Ba-133 and Am-241, both conditions displayed similar trendline respectively; the dry sample allowed relatively more gamma rays through wood compared to the wet sample. In the results for Ba-133, the wet and dry counts trendline were close to each other. The count difference between the thickest (position 7, 16.6 cm) and thinnest (position 1, 4.2 cm) through wood thickness is less obvious, i.e. dry 1678 counts, wet 2190 counts (Δ~512 counts). For Am-241, both wet and dry count trendlines are similar, but are further apart from each other. The wet sample having more attenuation due to the water absorbed in the wood after 24 hours soaking. The count difference between the thickest (position 7) and thinnest (position 1) wood thickness is more obvious, i.e. dry 18618 counts, wet 17885 counts (Δ733 counts). From the trendlines, Am-241 is the better choice in differentiating thickness, both in wet and dry conditions.

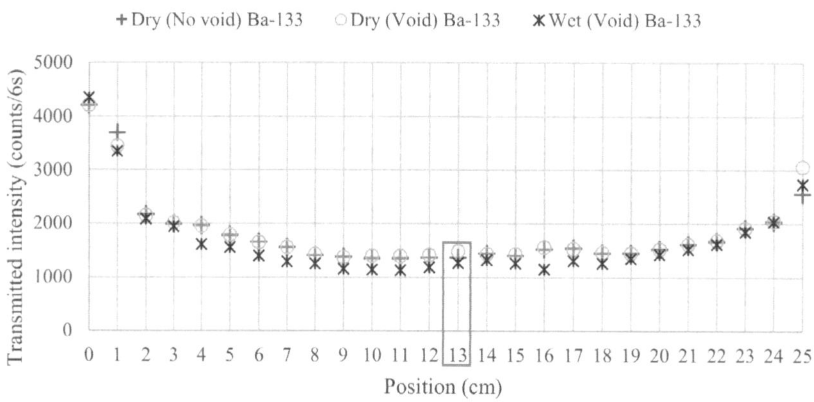

Figure 2. Density measurement: Wood Sample A -Khaya (branch) (Ba-133)

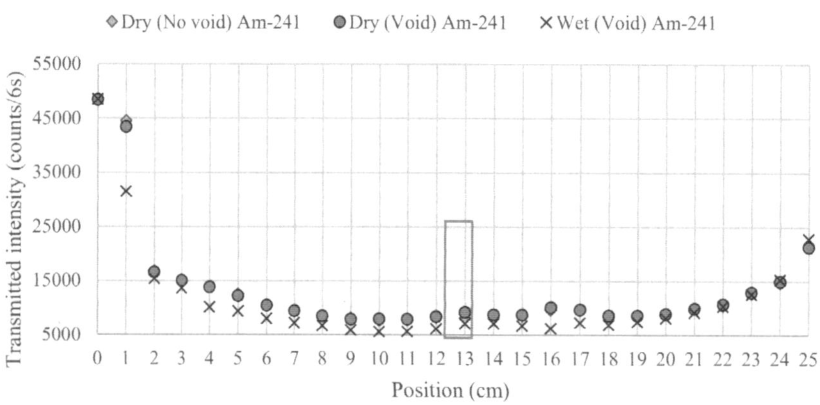

Figure 3. Density measurement: Wood Sample A- Khaya (branch) (Am-241)

In figure 4 at positions 8, 9, 10, 11 and 12 a void space is present. Void path lengths are 1.6cm, 2.9cm, 3.1cm, 2.5cm, 1.0cm respectively within the middle-area of the wood section. This void was created when the gaharu resin was taken out. We can see a slight uptrend of counts in this segment. The combined thickness (path length) of wood and void at positions 9 and 10 is 17.0cm, however their void sizes are 2.9cm and 3.1cm respectively. We see that this measurement is less sensitive to voids when it is dry compared to wet condition, especially when using Ba-133.

Figures 5 and 6 are results from density measurements of sample C (37.2 cm highest thickness) using Ba-133 and Am-241 respectively. Sample C has some rotting at the middle-segment of the wood, the void size is highest at positions 20 and 21 (void size 10.5cm). We see that in wet condition, both Ba-133 and Am-241 display mostly flat trendline compared to dry condition, and that gamma rays is further attenuated by the moisture within the wood.

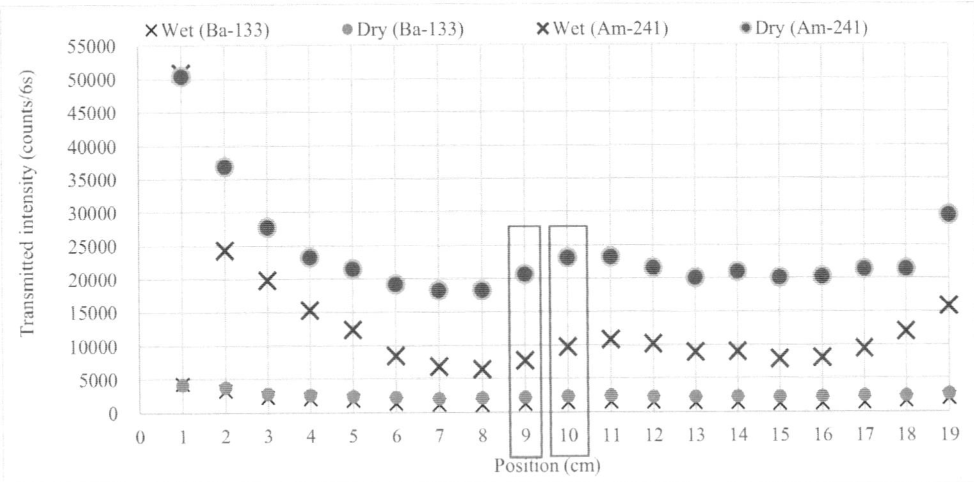

Figure 4. Density measurement: Wood Sample B Agarwood (Ba-133 and Am-241).

The presence of void at locations 15 through 28 with void sizes ranging between 2.0cm and 10.5cm is more obvious in the Am-241 results in figures 5 and 6, where we can see an upward drift at these locations. In Table 3, the attenuation coefficients for both wet and dry conditions is tabulated. This result supports the assertion that water content does provide additional attenuation, therefore suggest that if the measurements were done on-site at the plantation the moisture level would affect the data.

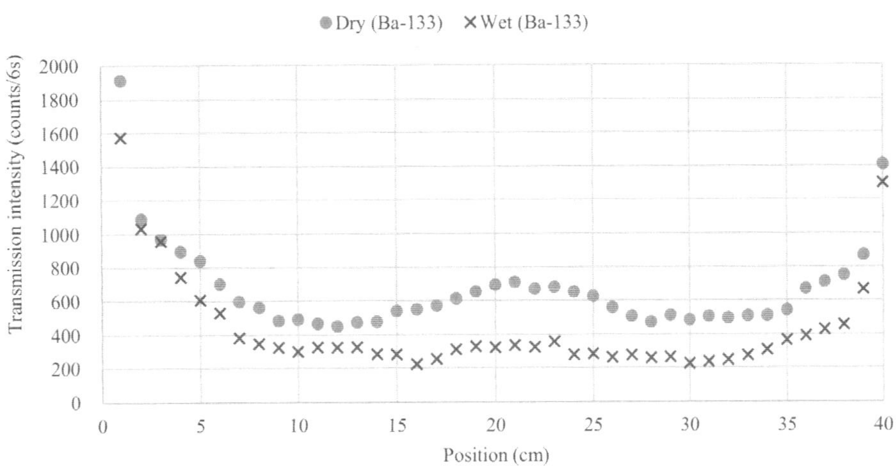

Figure 5. Density measurement: Wood C Khaya (trunk) (Ba-133).

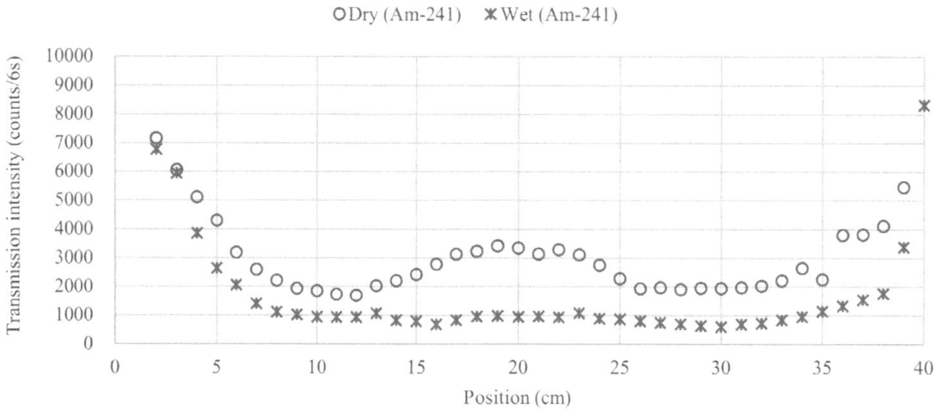

Figure 6. Density measurement: Wood C Khaya (trunk) (Am-241).

Table 3. Gamma attenuation coefficients for Agarwood and Khaya Senegalensis.

Sample	Mass Attenuation Coefficient (cm^2/g)		Diameter
	Am-241 (59.6keV)	Ba-133 (356keV)	(cm)
Agarwood (dry)	0.114±0.007	0.081±0.003	17.0 ±0.1
Agarwood (wet)	0.136±0.029	0.090±0.001	
Khaya-branch (dry)	0.136±0.004	0.086±0.004	24.0 ±0.1
Khaya-branch (wet)	0.141±0.004	0.088±0.004	
Khaya-trunk (dry)	0.120±0.004	0.068±0.028	37.2 ±0.1
Khaya-trunk (wet)	0.146±0.004	0.088±0.036	

4. Conclusion

Mass attenuation coefficient is a parameter that describes how much attenuation the incident ray undergoes by passing through a material. This parameter lends to quantitative analysis of a material. In density/thickness measurement and computed tomography, this coefficient is an important parameter for differentiating material composition and physical properties. This paper gives the estimated gamma ray attenuation coefficient for *Khaya Senegalensis*, a common shade tree planted for landscaping in Malaysia and Agarwood tree. Two radioisotopes, Am-241 and Ba-133 were utilised and results were compared to determine their suitability in wood section examination. It is found that for wood sections with densities between 0.40±0.01 g/cm^3 and 0.70±0.02 g/cm^3, Am-241 is more sensitive towards distinguishing thicknesses and densities compared to Ba-133. The results from the experiments can be a guideline for development of a gamma densitometer.

5. References

[1] Shakhreet B Z, Bauk S, Tajuddin A A and Shukri A 2009 Mass attenuation coefficients of natural Rhizophora spp. Wood for x-rays in the 15.77-25.27 keV range *Radiat. Protec. Dosim.* **135**(1) 47–53.
[2] Bradley D A, Tajuddin A A, Che Wan Sudin C W A and Bauk S 1991 Photon attenuation studies on tropical hardwoods. *Appl. Radiat. Isot.* **42**(8) 771-773.
[3] Boddupelli S and Nageshwara rao A S 2015 A study on photon interaction cross sections and mass attenuation coefficients dependents chemical composition of wood materials at the energy range of 1keV to 100MeV *Procedia Eng.* **127** 665-669.
[4] Saritha B and Rao A S N 2015 A study on photon attenuation coefficients of different wood materials with different densities *J.of Phys: Conf. Ser.* **662.**

[5] Ahmed J, Khan M M H, Kashem M A, Ahmed S H and Hossain M J 2019 Evaluation of density metric grading of agarwood, antioxidant potentially in agar oil and prevalence of unknown bacteria in agarwood soaking water. *J. Adv. Biotechnol. Exp. Ther.* **2**(2) 44-50 eISSN: 2616-4760.

[6] Gaitan-Alvarez J, Moya R and Berrocal A 2019 The use of X-ray densitometry to evaluate the wood density profile of Tectona grandis trees growing in fast-growth plantations *Dendrochronologia* **55** 71-79.

[7] Olale K, Yenesew A, Jamnadass R, Sila A and Sheperd K 2019 A simple field based method for rapid wood density estimation for selected tree species in Western Kenya *Scientific African* **5** e00149.

[8] Marashdeh M W, Al-Hamarneh I F, Munem E M A, Tajuddin A A, Ariffin A and Al-Omari S 2015 Determining the mass attenuation coefficient, effective atomic number, and electron density of raw wood and binderless particleboards of Rhizophora spp. by using Monte Carlo simulation. *Results in Physics* **5** pp228-234.

[9] Walker N K and Dodd R S 1988 Calculation of wood density variation from wood densitometer data. *Wood and Fiber Science* **20** (1) pp 35-43

[10] Ismail M P, Awang M R, Ahmad M N I, Abas A A and Osman M F **2008** Ultrasonic goniometry measurement of Aquilaria sp (gaharu) wood *Current Issues of Phys. in Malaysia* **1017**(1) AIP 978-0-7354-7/08/.

[11] Cown D J 2001 Wood Density in *Encylopedia of Materials: Science and Technology*, 2001.

[12] Nor Azah M A, Saidatul Husni S, Mailina J, Sahrim L, Abdul Majid J and Mohd Faridz Z 2013 Classification of agarwood (gaharu) by resin content *J. of Tropical Forest Sci.* **25(2)** 213-219.

[13] Gusmailina and Sumadiwangsa S 2020 The enhancement of added value of agarwood by diversification product *IOP Conf. Ser: Earth and Env. Sci.* **591** doi:10.1088/1755-1315/591/1/012022.

[14] Jok V A, Che Radzi N and Ku Hamid K H. 2015 A Review: Pharmacological properties of Aquilaria spp. Advanced Materials Research **1113** 193-197.

[15] Panduan Penanaman Pokok Teduhan. 2008. Jabatan Lanskap Negara, Kementerian Perumahan dan Kerajaan Tempatan.

[16] Yumi Zuhanis H -Y H, Nurhusna S, Anis Najiha A, Norsabrina M S, Gill C I R., Ummu Athiyah M T, Nur Iffah I and Nur Erina Syafiqa M R 2022 A bibliometric analysis of agarwood research 1959-2021 *J. of Agrobiotechnol.* **13**(2) 37-64.

[17] Zhang Y, Meng H, Lyu F, Fan X, Liu P, Huang Y, Chen B, Yang Y and Wei J 2023 Temporal characteristics of agarwood formation in Aquilaria sinesis after applying whole-tree agarwood-inducing technique **15** 37-44.

[18] Zhang N, Xue S, Song J *et al.* 2021 Effects of various artificial agarwood-induction techniques on the metabolome of *Aquilaria sinensis BMC Plant Biol* **21** 591 https://doi.org/10.1186/s12870-021-03378-8.

Depth dose measurement by using Al_2O_3 OSL dosimeters in high energy photons in the presence of air cavity and density inhomogeneity

Mohd Fahmi Mohd Yusof[1*], Amirah Sauki[1], Nur Ateqah Suzaini[1], Arifah Nazirah Abdullah[1], Ahmad Bazlie Abd Kadir[2], Norriza Mohd Isa[2]

[1]School of Health Sciences, Universiti Sains Malaysia, 16150 Kelantan, Malaysia
[2]Medical Physics and Secondary Standard Dosimetry Laboratory, Malaysian Nuclear Agency, 43000 Selangor, Malaysia

*mfahmi@usm.my

Abstract. Air cavities and tissue density inhomogeneity significantly affects the distribution of radiation doses, potentially resulting in adverse consequences such as cancer recurrence. This research aims to assess the accuracy of Al_2O_3 optically stimulated luminescence (OSL) dosimeters in measuring doses within varying thicknesses of air cavities (3, 5, and 8 cm) and tissue inhomogeneity of low and high density simulated by the lung and bone phantoms. An expanded polystyrene (EPS) was employed in homogeneous solid water® phantoms to simulate the air cavity. The percentage depth-dose (PDD) curves at 6 MV photons were obtained in both presence of air cavity and density inhomogeneity and compared to that in the EBT3 radiochromic film dosimeters and treatment planning system (TPS). The results indicated that the presence of an air cavity and tissue inhomogeneity affected the depth dose measured in OSL dosimeters, EBT3 films and TPS. OSLD and TPS showed good agreement at the centre of the cavity, which is within ±5% but could not estimate scattered radiation to the distal and proximal surfaces of the air cavity. The obtained p-values showed no significant differences of dose measured in OSL dosimeters compared to those in EBT3 films and TPS. The Kruskal Wallis test and Mann-Whitney showed no significant difference between OSL dosimeters, EBT3 film and TPS in the measurement of depth doses in the presence of density inhomogeneity. The overall results indicated the suitability of OSL dosimeters as indirect dosimeters for the measurements of depth dose in the presence of air cavity and tissue density inhomogeneity

1. Introduction

Dosimetry involves measuring and calculating the radiation dose received by a patient during treatment for ensuring accurate treatment dose delivery involving high energy photons. The ability of the dosimeter to accurately and efficiently measure the dose deposited in its effective volume is the main factor determining its applicability in dosimetry [1]. It is prominence to ensure the planned dose are accurately delivered to region of interest as it may increase the probability of secondary malignancy risk due to unnecessary doses delivered to the patients.

The human body consists of inhomogeneous density which refers to the high-density tissues such as bones and low-density tissue such as lung tissues [2]. The density differences have different physical and radiological characteristics giving significant impact of dose distributions. In addition, our body has a lot of heterogeneous interfaces such as lung-tissue and bone-tissue interfaces along the path of radiation beam causes increased or decreased transmission depending on the density and atomic number of the inhomogeneity medium [3].

The air cavities exist within the human body include the sinuses, the tracheobronchial tree, the oral cavity, the pharynx, and the nasal cavity. Additionally, the use of certain accessories like a mouth bite with an air tube for breathing purposes can also form large air cavities [4]. The presence of air cavities in the human body can significantly impact the distribution of high-energy radiation beams. This is due to the creation of electronic disequilibrium conditions near the interfaces between air and tissues, leading to insufficient doses at both the proximal and distal air cavity interfaces [5].

The Al_2O_3 optically stimulated luminescent (OSL) dosimeters have been extensively used as indirect dosimeter for the dosimetry works in radiotherapy and showed excellent agreement of dose readings involving high energy photons and electrons [6,7]. The current study focused on the dose measurements and to study the effectiveness of OSL dosimeters for dose measurements in in the medium on high energy photons with the presence of density inhomogeneity and air cavity. The OSL NanoDot®s was used and the depth doses in mediums with the presence of density inhomogeneity and air cavities were measured.

2. Methodology

2.1. Depth dose measurements in inhomogeneous density medium

The inhomogeneous density medium was prepared with the simulations of lung and bone tissues with the solid water phantoms. The lung was simulated by using cork slabs with average density of 0.24 g/cm^3 while the compact bone phantom with density of 1.85 g/cm^3 was used to simulate the bones. The lung and bone phantoms were sandwiched between the solid water phantoms to simulate the presence of low and high density mediums in solid water phantoms respectively as shown in Figure 1(a) and 1(b). The third set up of solid water, lung and bone phantoms are made to simulate the presence of all soft tissue, low and high density tissues at the same time as shown in Figure 1(c).The phantoms were irradiated to 6 MV photons with 300 cGy prescribed dose at 10 × 10 cm field size and 100 cm source to surface distance (SSD). Three sets of depth dose measurements were made on solid water-lung, solid water-bone and solid water-lung-bone setups with a bolus used to substitute solid water phantom at the locations of the OSL dosimeters to eliminate the air gaps. The depth doses were measured in several locations within the mediums including the interface regions (proximal and distal surfaces) and the interface between lung and bone.

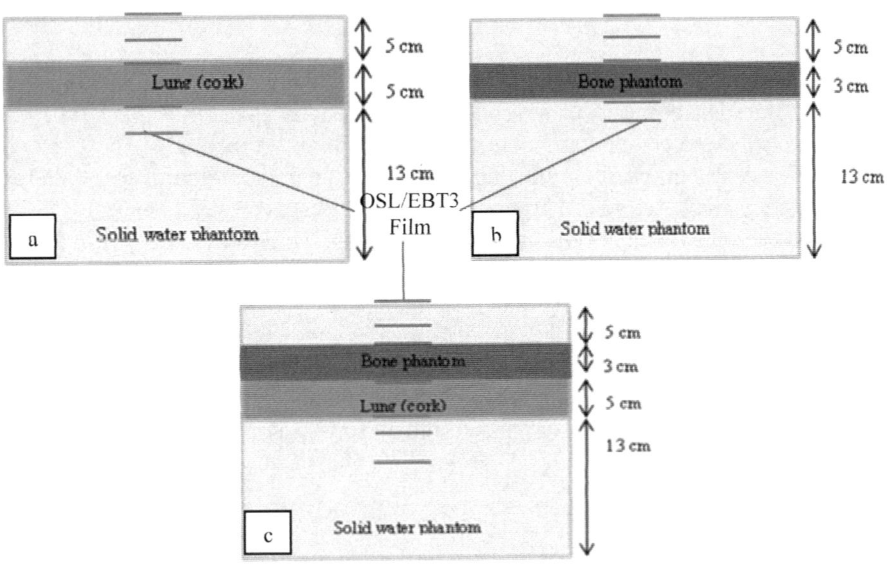

Figure 1. The preparations of inhomogeneous density phantoms of (a) solid water-lung phantom (b) solid water-bone phantom and (c) solid water-lung phantom-bone phantom.

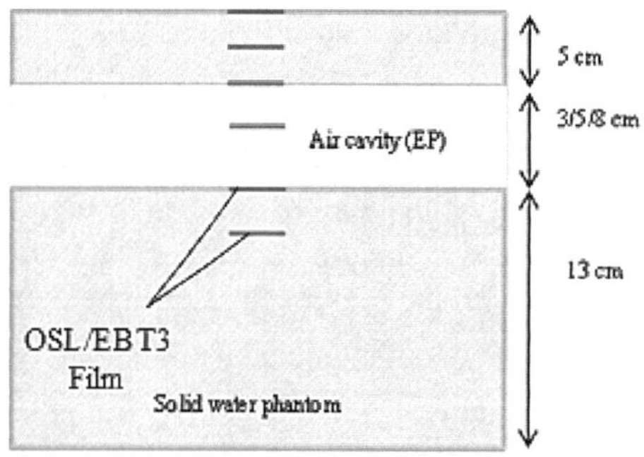

Figure 2. The phantom setup for the presence of air cavity simulated by using extended polystyrene (EP) with thicknesses between 3 and 8 cm.

2.2. Depth dose measurement in the presence of air cavity
An extended polystyrene (EP) with the density of 0.01 g/cm3 was used to simulate the air cavity [8]. The EP was sandwiched between the solid water phantoms with various thicknesses between 3and 8 cm as shown in Figure 2. The air cavities were simulated at thicknesses of 3, 5 and 8 cm. The phantoms were irradiated to 6 MV photons with 300 cGy prescribed dose at 10×10 cm field size and 100 cm source to surface distance (SSD). The depth doses were measured in several locations within the mediums including the interface regions (proximal and distal surfaces) between solid water phantoms and EP as well as within the EP.

2.3. Simulation by using treatment planning system
The phantoms of inhomogeneous density and the presence of air cavities were scanned by using a computed tomography (CT) scanner and determined of their depth doses by using a treatment planning system (TPS). The depth doses of the similar points to the OSL dosimeters were determined by using the Oncentra 4.3 TPS software used for the treatment planning for radiotherapy. All the depth doses measured in the OSL dosimeters were plotted and compared to those in the EBT3 film dosimeters and TPS.

3. Results and discussion
Figure 3 illustrated the PDD curve by using OSL dosimeters in comparison to the EBT3 film and TPS in solid water-lung, solid water bone and solid water-bone-lung respectively. The presence of lower and higher density tissue simulated by the lung and bone phantoms significantly altered the PDD curve in 6 MV photons compared to the homogeneous density solid water phantoms [9,10]. The dose measured by the OSL dosimeters were in good agreement with the EBT3 film dosimeters with maximum percentage differences within 2.5% in all similar depths of measurements. The density inhomogeneity did not significantly affect the depth doses in TPS compared to the homogenous density phantoms. The depth doses by the OSL dosimeters were significantly lower in the lung-bone mediums compared to the EBT3 films although the two dosimeters showed similar trend of dose perturbations. Table 1-3 summarized the depth dose measured by the OSL dosimeters, EBT3 film dosimeters and TPS of the similar depths and locations in solid water-lung, solid water-bone and solid water-bone-lung phantoms.

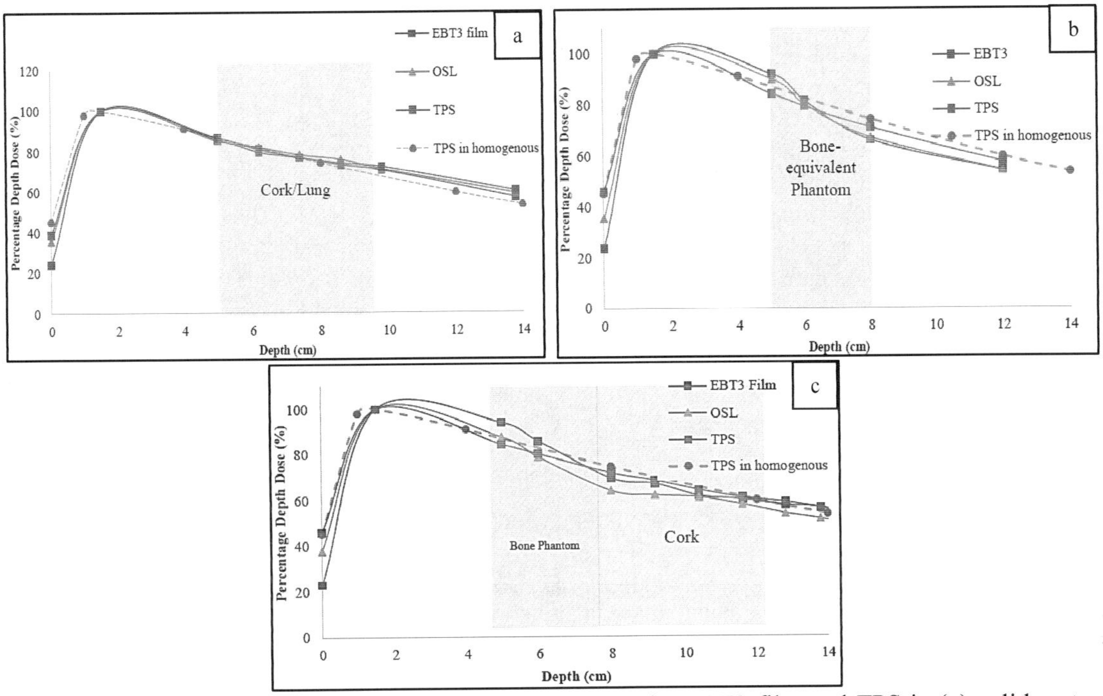

Figure 3. The PDD in OSL dosimeters in comparison to the EBT3 film and TPS in (a) solid water-lung (b) solid water-bone and (c) solid water-lung-bone.

Table 1. The depth dose measured in the OSL dosimeters, EBT3 film dosimeters and TPS at the similar depths in solid water-lung phantoms.

Depth (cm)	Description	Percentage depth dose (%)		
		OSL	EBT3 Film	TPS
0	Surface	35.79	24.33	38.81
5.0	Upper solid water-lung interface	85.25	84.25	86.74
7.4	Centre of lung	78.25	77.25	77.01
9.8	Lower solid water-lung interface	71.04	71.4	72.21

Table 2. The depth dose measured in the OSL dosimeters, EBT3 film dosimeters and TPS at the similar depths in solid water-bone phantoms.

Depth (cm)	Description	Percentage depth dose (%)		
		OSL	EBT3 Film	TPS
0	Surface	35.38	45.85	45.85
5.0	Upper solid water-bone interface	89.97	91.94	84.18
7.4	Centre of bone	79.96	81.46	79.22
9.8	Lower solid water-bone interface	67.04	66.02	67.04

Table 3. The depth dose measured in the OSL dosimeters, EBT3 film dosimeters and TPS at the similar depths in solid water-bone-lung phantoms.

Depth (cm)	Description	Percentage depth dose (%)		
		OSL	EBT3 Film	TPS
0	Surface	37.89	22.05	46.01
5	Upper solid water-bone interface	87.66	94.05	84.38
6	Centre of bone	75.59	85.76	79.98
8	Bone-lung interface	63.74	69.20	71.56
10	Centre of lung	60.77	61.38	63.93
13	Lung-solid water interface	53.65	57.24	58.62

Figure 4 illustrated the PDD curve of OSL dosimeters in comparison to EBT3 film and TPS in in the presence of air cavities of 3, 5 and 8 cm thicknesses. The presence of air cavities significantly reduced the depth dose in all air cavity thicknesses [11]. The depth dose in OSL dosimeters showed good agreement with EBT3 film dosimeters at all air cavity thicknesses. The depth dose in OSL dosimeters however were overestimated compared to the EBT with the maximum percentage of discrepancies of 15.46% when the thickness of air cavities increased. This is due to the depth dose measured in EBT3 films declined more significant compared to that in OSL dosimeters when the air cavities increased. It was observed also that the depth dose measured in OSL and EBT3 film dosimeters were not in agreement to the TPS declining of depth dose in TPS becoming less at increased air cavity thicknesses. Table 4 summarized the percentage dose measured by using OSL dosimeters, EBT3 films and TPS at similar depths with the presence of air cavities with thicknesses of 3, 5 and 8 cm.

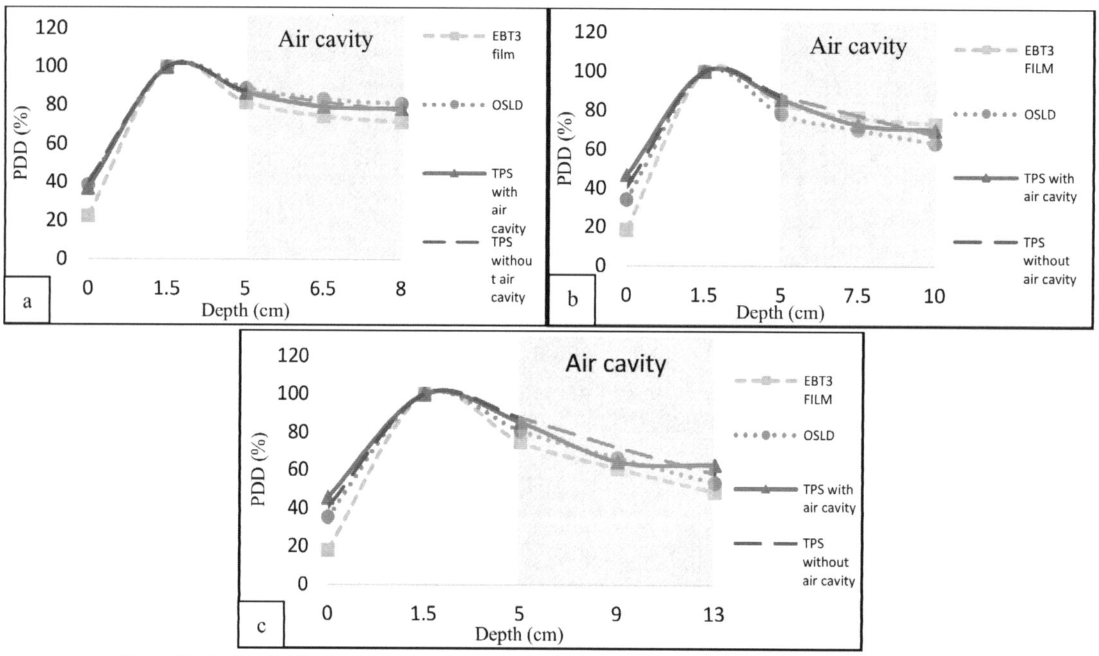

Figure 4. The PDD in OSL dosimeters in comparison to the EBT3 film and TPS in solid water-air cavity of (a) 3 cm (b) 5 cm and (c) 8 cm thicknesses.

Table 4. The depth dose measured in the OSL dosimeters, EBT3 film dosimeters and TPS at the similar depths in the presence of air cavities of different thicknesses.

Air cavity thickness (cm)	Position	Percentage depth dose (%)		
		OSL	EBT3 Film	TPS
3 cm	Surface	38.50	22.90	36.90
	Upper solid water-air interface	88.90	81.55	86.64
	Centre of air cavity	83.37	74.73	79.45
	Lower solid water-air interface	80.91	71.39	78.42
5 cm	Surface	34.1	19.71	46.58
	Upper solid water-air interface	78.16	82.37	82.37
	Centre of air cavity	70.13	74.35	74.35
	Lower solid water-air interface	63.20	68.70	69.70
8 cm	Surface	35.3	18.60	45.89
	Upper solid water-air interface	80.83	74.86	85.62
	Centre of air cavity	66.57	61.02	64.73
	Lower solid water-air interface	53.70	48.97	63.36

4. Conclusion

The measurement of PDD showed good agreement of dose reading by using OSL dosimeters to EBT3 film dosimeters in inhomogeneous density medium and the presence of air cavity. The thickness of air cavities did not significantly change the sensitivity of OSL dosimeters. The overall results indicated the suitability of OSL NanoDots® to be used as inactive or indirect dosimeter for measuring the dose in mediums in the presence of inhomogeneous densities and air cavity. The overestimation of dosimetry by OSL dosimeters to the EBT3 films in air cavities shall be investigated in the future work to understand the dosimety behavior of OSL for in-air dosimetry.

5. References

[1] Izewska, J and Rajan G 2005 *Radiat. Oncol. Phys: A Handbook for Teachers and Students*, (International Atomic Energy Agency-IAEA, Vienna).

[2] Zaman A, Kakakhel M B, and Hussain A 2019 A comparison of Monte Carlo, anisotropic analytical algorithm (AAA) and Acuros XB algorithms in assessing dosimetric perturbations during enhanced dynamic wedged radiotherapy deliveries in heterogeneous media *J. of Radioth. in Prac.* **18**(1) 70–74.

[3] Verma R T, Painuly K N, Mishra P S, Yoganathan S A, Singh N, Bhatt M L B and Jamal N 2019 Evaluation of lung density and its dosimetric impact on lung cancer radiotherapy: A simulation study *J. Biomed. Phys. Eng.* **9**(1) 17-28.

[4] Joshi C P, Darko J, Vidyasagar P B, and Schreiner L J 2010 Dosimetry of interface region near closed air cavities for Co-60, 6 MV and 15 MV photon beams using Monte Carlo simulations *J. Med. Phys.* **35**(2) 73–80.

[5] AAPM Tissue inhomogeneity corrections for megavoltage photon beams Report No. 85 2004, (Madison: American Association of Physics in Medicine, Medical Physics Publishing).

[6] Yusof M F M, Othman N N M, Abdullah R, Kadir A B A and Salleh H 2020 Percentage depth dose measurement in high energy photons and electrons by using the Al2O3 optically stimulated luminescent (OSL) dosimeter *IOP Conf. Series: Mater. Sci. Eng.* **785**(1), 012002.

[7] Yusof M F M, Shariff N A, Fauzi N S M, Kadir A B A, and Isa N M 2022 Small beam dosimetry by using Al_2O_3 optically stimulated luminescent (OSL) dosimeters at high energy photons and electrons *IOP Conf. Series: Mater. Sci. Eng.* **1231**(1), 012018.

[8] Song Y K, Hong S H, Eo S, Han M G and Shim W J 2020 Rapid production of micro- and nanoplastics by fragmentation of expanded polystyrene exposed to sunlight *Environ. Sci. Technol.* **54**(18), 11191–11200.

[9] Khan M F and Gibbons J 2014 *Khan's Physics of Radiation Therapy 5th Ed.* (Philadelphia Lippincott William & Wilkins).

[10] Woon Y L, Heng S P, Wong J H D and Ung NM 2016 Comparison of selected dose calculation algorithms in radiotherapy treatment planning for tissues with inhomogeneities. *J. Phys.: Conf. Ser.* **694** 012024.

[11] Sabater S, Berenguer R, Honrubia-Gomez P, Rivera M, Nunez A, Jimenez-Jimenez E, Martos A and Ramirez-Castillo C 2014 How air influences radiation dose deposition in multiwell culture plates: a Monte Carlo simulation of radiation geometry. *J. of Radiat. Res.* **2014** (1-6).

Acknowledgement

The authors would like to acknowledge the financial support from the USM Research Universiti (RU-I) grant scheme no 1001.PPSK.8012381. The authors also would like to acknowledge the Nuclear Medicine, Radiotherapy and Oncology Department, Hospital Universiti Sains Malaysia (HUSM) for the use of CT scanner and TPS software in this study.

Dose verifications in several organs in abdominal computed tomography (CT) scan by using Al_2O_3 OSL dosimeters

Mohd Fahmi Mohd Yusof[1*], Muhammad Haiqal Ramli[1], Che Nazri Che Hussin[1], Ahmad Bazlie Abd Kadir[2], Norriza Mohd Isa[2]

[1]School of Health Sciences, Universiti Sains Malaysia, 16150 Kelantan, Malaysia
[2]2Medical Physics and Secondary Standard Dosimetry Laboratory, Malaysian Nuclear Agency, 43000 Selangor, Malaysia

*mfahmi@usm.my

Abstract. The study investigated the dose of several organs in the computed tomography (CT) scan by using Al_2O_3 optically stimulated luminescent (OSL) dosimeters. The organs of right and left kidney, right pancreas, spleen and liver in an anthropomorphic human phantom were investigated for the dose received during the scan. The OSL dosimeters are embedded into custom-made wax and replace onto two selected slices of human body phantom where the selected organs are located. The phantom was scanned with by using adult abdomen protocol. The percentage differences of OSL doses reading compared to the previous study data and the international dose reference level (DRL). The doses obtained from OSL NanoDots® showed excellent agreement compared to XRQA film dosimeters with percentage differences within 5%. The doses measured by using OSL dosimeters in five organs were also within the range of the national DRL values by the Ministry of Health of Malaysia (MOH). These findings suggest that our results can be utilized to verify the doses received by internal organs during computed tomography abdomen protocols. The overall results indicated the suitability of OSL dosimeters for the indirect dose verification in the CT scan.

1. Introduction

Due to its extensive use, the computed tomography (CT) became the leading contributor to the collective effective dose among all diagnostic imaging procedures. It has therefore increased the awareness of the potential radiation risks to the population [1]. CT would contribute up to 70% of the total dose given to patients [2]. The National Commission of Radiation Protection (NCRP) Report 160 indicated that the medical exposure as the highest contributor to the man-made radiation exposure with CT scan as the greatest contributor at 49% of medical exposure [3]. The major risks associated with the use of CT include the possibility of cancer induction from x-ray radiation exposure. Therefore, the call for accurate dose verification receive by patients from standard CT protocol used in clinical examination is required to ensure the safety of the patients undergoing CT scans.

The International Commission on Radiological Protection (ICRP) introduced the diagnostic reference levels in 19901 in 1996 as a method to determine the patient's dose at the safe levels [4]. Diagnostic reference levels are measures of common practise in a country or region to develop national or regional diagnostic reference values due to different imaging protocols in various nations. DRLs are determined by radiation protection and health authorities along with professional medical organisations of the nation and often used to reduce the patient dose in CT scans while optimizing the quality of the radiographic images [5].

The Al_2O_3 Optically stimulated luminescence (OSL) dosimeters are typically used for clinical dosimetry in many medical physics works due to its size as an in vivo dosimeter with good spatial resolution and wide dose range [6,7]. Previous study suggested the suitability of the OSL dosimeters for dosimetry works in diagnostic kilovoltage x-rays and investigations of organ doses in general radiography [8,9] The OSL dosimeters also showed good agreement of dose to the CT dose index (CTDI) and dose mapping in CT scans [10,1,11]. This study focused on the measurement of dose of several organs (kidney, spleen, pancreas and liver) by utilise the OSL dosimeters in comparison with the available DRL. The OSL NanoDot® in this study was used to investigate the dose received by several critical organs during the CT abdominal scanning.

2. Methodology

2.1. Energy calibration of OSL dosimeters
OSL dosimeters are made of aluminium oxide (Al_2O_3) having a relatively high effective atomic number (Z_{eff} = 11.28). This caused the material to exhibit an exaggerated response of low-energy photons that usually use in diagnostic radiology due to the significant contribution of the photoelectric effect [12]. The OSL dosimeters were calibrated by the secondary standard dosimetry laboratory (SSDL), Malaysian Nuclear Agency by using x-rays of 80 kVp energy to set up the dose analysis method by using the OSL reader. The OSL dosimeters were then calibrated with x-ray energies between 80 and 10 kVp to determine their energy dependency on kilovoltage x-ray

2.2. Measurement of organ dose in CT abdominal scan
An anthropomorphic human phantom was used to simulate the human anatomical features as shown in Figure 1(a). One slab of the phantom was replaced with paraffin wax for the positioning of the OSL dosimeters as shown in Figure 1(b) [9]. This is due to unavailability to position the OSL dosimeters within the slabs of the anthropomorphic phantom. The paraffin wax was shaped based on the replaced slice of the anthropomorphic phantom and the OSL dosimeters were placed at the positions of the critical organs of right kidney, left kidney, pancreas, spleen and liver. The anthropomorphic phantom was scanned by using a CT scanner at abdominal scanning protocol (120 kVp and 250 mAs). The exposed OSL dosimeters were read by using the OSL reader and compared to the dose measured in XRQA film dosimeter and available DRL [13].

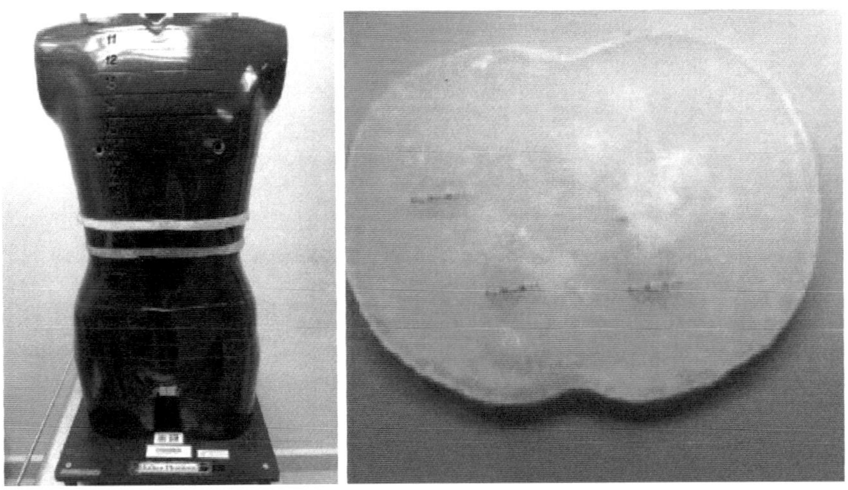

Figure 1. (a) The anthropomorphic human phantom used to simulate human body and (b) the paraffin wax used to replace the phantom slice consisting the critical organs.

3. Results and discussion
Figure 2 showed the dose response of the OSL dosimeters towards the increased x-ray energies between 80 and 140 kVp. The dose recorded by the OSL showed excellent agreement to the increased

x-ray energies shown by the linear regression values, R^2 near to 1.0. This indicated good energy dependent and sensitivity of the OSL dosimeters towards the increase of x-ray energies [14,15].

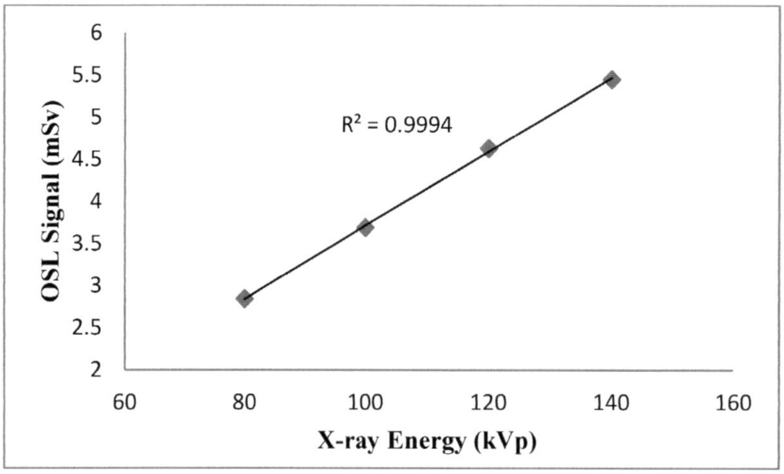

Figure 2. The dose response curve of OSL dosimeter at x-ray energies between 80 and 140 kVp.

Table 1 showed the dose comparison between the OSL dosimeters and XRQA film dosimeters. The Dose measured by using OSL dosimeters showed excellent agreement to the XRQA film dosimeter at all measured organs with maximum percentage difference lower than 5%. The results are also in good agreement with the previous work by Brady et al., [13] that compared the dose measured in OSL dosimeters and XRQA film dosimeters in both XVI and OBI cone beam systems in the CT scan.

Table 1. The dose measured in several organs by using OSL dosimeters and XRQA film dosimeter.

Organ	Dose in OSL[a] (mGy)	Dose in XRQA film[b] (mGy)	Percentage difference (%)
Right kidney	25.21	25.26	0.20
Left kidney	24.35	23.24	4.66
Liver	27.83	27.28	1.99
Pancreas	28.02	27.89	0.47
Spleen	24.17	24.78	2.49

[a]Current study
[b]Brady et al., (2010)

Table 2 showed the dose measured in the critical organs by using the OSL dosimeters during the abdominal CT scanning in comparison with the national DRL and other DRL [16,17]. The national DRL did not provide the specific value for individual organs and only gives the dose range for all organs. The Slovenian DRL was referred in this study to as it provides the DRL for individual organs to be compared with the dose readings obtained from the current study. The doses measured by using the OSL dosimeters in this study showed high percentage differences between 31.3 and 38.99% compared to the DRL [17]. The dose readings in the OSL dosimeters in this study however showed good consistency of readings within the organs being measured. Kawaguchi et al., [15], suggested that the use of OSL dosimeters for very low energy x-rays such as mammography require adjustments for energy dependence and this could also applied for the CT x-ray energies.

Table 2. The dose measured in several organs by using OSL dosimeters in comparison to the dose reference levels (DRL).

Organ	Dose in OSL (mSv)	Dose Reference Level (DRL[a]) (mSv)	Dose Reference Level (DRL[b]) (mSv)	Percentage difference to DRL[b] (%)
Right kidney	25.21		17.98	33.48
Left kidney	24.35		17.76	31.30
Liver	27.83	12.8 – 39.1	18.75	38.99
Pancreas	28.02		19.52	35.76
Spleen	24.17		17.50	32.01

[a]Current study
[b]Zalokar et al., (2020)

4. Conclusion

The doses obtained by OSL nanodots showed excellent agreement when compared to the doses obtained by using XRQA films dosimeter. The values of doses assessed in 5 organs undergo CT abdomen protocol are within the range of national DRL values from Ministry of Health Malaysia (MOH). This indicates our results can be used to verify the doses receive by internal organs in computed tomography abdomen protocol. The use of DRL should be in regional level as difference country would have difference value of their national DRL. The overall results indicated the suitability of OSL dosimeters for in-vivo dosimetry works in CT scans. There are also needs for specific DRL values for organs in the national level as part of quality assurance and radiation protection to the patients.

5. References

[1] Zhang D, Li X, Gao Y, Xu X P, and Liu B 2013 A method to acquire CT organ dose map using OSL dosimeters and ATOM anthropomorphic phantoms *Med. Phys.,* **40**(8), 081918–081918.

[2] Corona C E, Ferreira I-B G, Herrera, J G, López S R, and Covarrubias O S 2015 Verification of CTDI and DLP values for a head tomography reported by the manufacturers of the CT scanners, using a CT dose profiler probe, a head phantom and a piranha electrometer. *Int. Symp.on Solid State Dosimetry* 2015 (International Atomic Energy Agency (IAEA)).

[3] Bolus N E 2013 NCRP Report 160 and what It means for medical imaging and nuclear medicine *J. Nucl. Med. Technol.* **41**(4) 255–260.

[4] McCollough C H, Branham T, Herlihy V, Bhargavan M, Robbins L, Bush K L, McNitt-Gray M F, PayneJ T, Ruckdeschel T, Pfeiffer D G, Cody D D and Zeman R K 2011 Diagnostic Reference Levels From the ACR CT Accreditation Program *J. Am. Coll. Radiol.* **8**(11), 795–803.

[5] Muhammad N A, Karim M K A, Hassan H A, Kamarudin M A, Wong J H D and Hoong N K 2020 Diagnostic Reference level of radiation dose and image quality among paediatric CT Examinations in a tertiary hospital in Malaysia *Diagnostics* **10**(8), 591.

[6] Yusuf M, Alothmany N and Kinsara A A 2017 Organ dose measurement using optically stimulated luminescence detector (OSLD) during CT examination *Radiat. Phys Chem.,* **139** 83-89.

[7] Ponmalar Y, Manickam R, Saminathan S and Ganech K M, 2017 Response of nanodot optically stimulated luminescence dosimeters to therapeutic electron beams *J. Med. Phys.* **42**(1) 42.

[8] Radzi M N S W M 2022 *Dosimetry Characteristics of Optically Stimulated Luminescent (OSL) Dosimeters in kilovoltage energy range* B.Sc Thesis (Universiti Sains Malaysia).

[9] Zulkarnain A 2022 *Verification of impact of anode heel orientation to radiosensitive organ and effective dose in AP pelvis radiography using OSL Nanodot.* B.Sc Thesis (Universiti Sains Malaysia).

[10] Ruan C, Yukihara E G, Clouse W J, Gasparian P B R and Ahmad S 2010 Determination of multislice computed tomography dose index (CTDI) using optically stimulated luminescence technology *Med. Phys.* **37**(7) 3560–3568.

[11] Giansante L, Santos J C, Umisedo N K, Terini 2018 Characterization of OSL dosimeters for use in dose assessment in computed tomography procedures *Phys. Med.* **47** 16–22.

[12] Al-Senan R and Hatab M R 2011 Characteristics of an OSLD in the diagnostic energy range. *Med. Phy.* **38**(7) 4396–4405.

[13] Brady S, Yoshizumi T, Toncheva G and Frush D 2010 Implementation of radiochromic film dosimetry protocol for volumetric dose assessments to various organs during diagnostic CT procedures. *Med. Phys.* **37**(9) 4782–4792.

[14] Yusof M F M, Yahya M H, Rosnan M S, Abdullah R and Kadir 2017 Dose measurement using Al2O3 dosimeter in comparison to LiF:Mg,Ti dosimeter and ionization chamber at low and high energy x-ray *AIP Conf. Proc.* **1799**, 040007.

[15] Kawaguchi A, Matsunaga Y, Suzuki S and Chida K 2017 Energy dependence and angular dependence of an optically stimulated luminescence dosimeter in the mammography energy range *J. Appl. Clin. Med. Phys.* **18**(2) 191–196.

[16] Ministry of Health Malaysia 2013 *Malaysian Diag.. Ref. Levels in Med. Imaging (Radiol.).*

[17] Zalokar N, Marciuš V Z, and Mekiš N 2020 Establishment of national diagnostic reference levels for radiotherapy computed tomography simulation procedures in Slovenia *Eur. J. Radiol.* **127** (108979).

Acknowledgement

The authors would like to acknowledge the financial support from the USM Research Universiti (RU-I) grant scheme no 1001.PPSK.8012381. The authors also would like to acknowledge the Radiology Department, Hospital Universiti Sains Malaysia (HUSM) for the use of CT scanner in this study.

IOP Conf. Series: Materials Science and Engineering 1308 (2024) 012013 doi:10.1088/1757-899X/1308/1/012013

Subsurface utility detection using ground penetrating radar and electromagnetic locator – a comparative study

Amer Hazreq bin Haron[1,a)], Tengku Sarah binti Tengku Amran[1], Mohamad Ridzuan Bin Ahmad[1], Noor Azreen Masenwat[1], Suhairy Sani[1], Ismail Mustapha[1], Shaharudin Sayuti[1], Noryana Binti Abdul Razak[1], Mior Ahmad Khusaini Bin Adnan[1], Muhamad Ali bin Mohd Yusoff[2], Muhammad Aliff Daniel bin Mohsin[2]

[1]Material Structure Integrity Group (NDT-MSI), Industrial Technology Division, Malaysian Nuclear Agency, 43000, Kajang, Selangor, Malaysia.
[2]Electrical Engineering Technology Faculty, Kampus Alam Pauh Putra, Universiti Malaysia Perlis (UniMAP), 02600, Arau, Perlis, Malaysia.

[a] amer@nm.gov.my

Abstract. The process of finding and mapping underground utilities is crucial for managing underground infrastructure effectively, especially to avoid damaging them during construction and digging. This research studies two main methods used for this purpose: Ground Penetrating Radar (GPR) and Electromagnetic Locator (EML). GPR works by using electromagnetic waves to detect and assess things underground, giving an idea of how deep they are. On the other hand, EM L help to find metal utilities by using electromagnetic fields. This shows how important it is to have accurate maps of these utilities. Having precise maps is like having a solid foundation for planning and maintaining infrastructure. It also points out that when combine information from GPR and EML and analyse the results together, it can make the findings even more accurate. In a nutshell, these methods are crucial for keeping us safe, preventing delays in projects, and making sure to use resources wisely across different industries. In this research, Root Mean Square Error (RMSE) is used to determine the correlation in depth accuracy error which is 0.05821 which is calculated based on the depth measured by GPR and EML.

1. Introduction

Utility location involves the process of identifying and labelling underground public utility mains, which encompass various types such as telecommunication lines, electricity distribution, natural gas pipelines, fibre optics and wastewater pipes. Preserving the accuracy and reliability of underground utility mapping and its database is crucial for planning, upgrading, and maintaining infrastructure, particularly below the ground. This mapping encompasses the positioning, location, and identification of buried infrastructure and objects, including archaeological artefacts. Utility detection and mapping applications extend to areas like metallic and non-metallic utilities, environmental assessment, damage prevention, geological investigation, turf assessment, archaeology, forensics, and road inspections[1].

In today's modern world, there's a growing demand for services delivered through underground pipes and cables, collectively referred to as utilities. These services include telecommunication, internet access, water supply, and sewage removal for both residential and commercial purposes. As a result, a complex network of pipes and cables has emerged underground, varying in depth and direction. Just as space is limited for buildings above ground in cities, the underground space also faces limitations for access and accommodating utilities. GPR is an effective geophysical technique for investigating buried sites, especially when features of interest lie within 2-3 meters beneath the surface. Archaeologists increasingly use GPR for landscape analysis, as it provides maps and images guiding excavation placement and offers a broader context for studying human interaction with ancient landscapes.

Various methods and equipment, such as GPR, EML, seismic waves, excavation, and x-ray scanning, are employed to identify buried utilities, chosen based on project requirements and cost considerations. To prevent accidents and damage during excavation, utility maps displaying as-built infrastructure are essential references. Consequently, utility mapping has become an integral part of the land survey profession, ensuring sustainable development by safeguarding underground utilities from harm[2].

Utility detection and correlation using GPR and EML are essential techniques in industries like construction and utilities management. Both GPR and EML are often used together to cross-verify and correlate findings, providing more accurate data for safe and efficient excavation and construction, ultimately reducing project risks and costs. However, their effectiveness can vary based on factors such as soil conditions and operator expertise, so it's crucial to follow local regulations and consult professionals when employing these technologies [3].

2. Background

Mapping underground systems, such as pipes and cables, is essential across various industries to support effective construction, upkeep, and repair activities. Two widely employed methods for underground utility mapping are EML and GPR. EML devices employ electromagnetic induction for detecting buried cables and pipes, whereas GPR relies on radar waves to generate images of the subsurface. Researchers Siu and Lai [4] undertook a study to assess the precision of underground utility locations using EML (RD8100) and GPR (Mala ProEx). Their findings indicated that GPR outperformed EML in terms of accuracy.

Furthermore, another investigation by Siu and Lai [4] centred on the precision of depth measurements obtained with GPR (Mala ProEx) and EML (RD8000). The research highlighted the significance of taking into account variables like depth, location, material composition, geological conditions, and proximity to other utilities when applying these methods. Additionally, Siu and Lai conducted a study [4] that compared the precision of GPR (Mala ProEx) and EML (RD8000) in various types of utility lines.

Investigating the relationship between EML and GPR for underground mapping is a significant research area. Grasping the advantages and constraints of each method can assist industry professionals in selecting the most appropriate approach tailored to their specific underground mapping requirements. To sum up, numerous studies have delved into the connection between EML and GPR for mapping underground systems. Although EML devices provide convenience, reliability, and ease of operation, GPR devices excel in achieving greater accuracy when pinpointing underground utilities. Additional research and in-depth analysis are required to gain a comprehensive understanding of the relationship between these two methods and to enhance their synergistic utilization in underground mapping.

2.1. Ground penetrating radar

GPR is a geophysical technique that uses electromagnetic waves to image and detect subsurface features and objects. It has become a widely used tool in various fields, including archaeology, geology, civil engineering, and utility detection. This literature review aims to provide an extensive overview of the applications, methodologies, and advancements in GPR technology.

2.1.1. GPR applications. GPR has revolutionized archaeological investigations by providing non-destructive imaging of buried structures, artefacts, and geological features. Conyers [5] discusses the use of GPR in garden archaeology, where it has been used to map buried features, identify ancient pathways, and locate potential excavation sites. GPR has also been used to study ancient landscapes and human interaction with the environment [5].

GPR plays a crucial role in civil engineering projects, including infrastructure assessment, pavement evaluation, and utility detection. It can provide information about subsurface conditions, such as the thickness of pavement layers, the presence of voids, and the location of buried utilities. Dusan and Aleksandar [6] highlight the use of GPR and GPS technology for the detection of underground utilities, emphasizing its accuracy and efficiency in utility mapping.

GPR is widely used in geology and environmental studies to investigate subsurface structures, soil properties, and groundwater resources. It can help identify geological formations, map bedrock depths, and detect groundwater levels. Jol [7] provides a comprehensive overview of GPR theory and applications in geology, discussing its use in mapping geological structures, characterizing aquifers, and studying soil properties.

2.1.2. GPR methodologies. Data Acquisition: GPR data is acquired by transmitting electromagnetic waves into the ground and recording the reflections from subsurface features. The data is collected using antennas or transducers that emit and receive electromagnetic signals. The choice of antenna frequency depends on the desired depth of investigation and the resolution required. Higher frequencies provide better resolution but have limited penetration depth, while lower frequencies have greater penetration but lower resolution [7].

Data Processing and Interpretation: GPR data processing involves filtering, stacking, and migration techniques to enhance the quality and clarity of the subsurface images. Various software tools, such as RADAN 7, are used for data analysis and interpretation.

2.2. Electromagnetic locator

EML are essential tools used in various industries for accurately detecting and mapping underground utilities such as pipes and cables. This literature review aims to provide an extensive overview of the applications, methodologies, and advancements in EML technology.

2.2.1. EML applications. EMLs are widely used in construction and infrastructure projects to locate and map underground utilities before excavation. This helps prevent accidental damage to utility lines, ensuring worker safety and minimizing costly repairs. EMLs are also used for mapping utility networks during the planning and design stages of construction projects [8].

EMLs play a crucial role in maintenance and repair operations, allowing technicians to quickly and accurately locate faults or breaks in underground utility lines. This enables efficient repairs and reduces downtime for utilities such as water, gas, and electricity [5].

Telecommunications: EMLs are extensively used in the telecommunications industry to locate and trace underground cables. This is essential for installing new cables, repairing damaged lines, and identifying cable routes for network expansion. EMLs help ensure the efficient and reliable operation of telecommunications networks [6].

2.2.2. EML methodologies. Electromagnetic Induction: EMLs use electromagnetic induction principles to detect and locate underground utilities. They emit electromagnetic signals into the ground and measure the response from metallic objects, such as pipes and cables. The strength and frequency of the signals can be adjusted to optimize detection accuracy and depth penetration [6].

Signal Detection and Interpretation: EMLs employ various signal detection and interpretation techniques to accurately locate underground utilities. These techniques include peak detection, signal strength analysis, and signal filtering to distinguish between different types of utilities and minimize interference from nearby objects [7].

3. Methodology

The methodology for utility detection and correlation plays a vital role in ensuring the safety and efficiency of construction and excavation projects. Key methods in this process are GPR and EML. These techniques help identify and map subsurface utilities such as water pipes, gas lines, and electrical cables with precision. Using GPR and EML effectively can prevent costly and hazardous utility damage. This methodology emphasizes the importance of accurate utility mapping and the advantages of combining GPR and EML findings for improved results.

The site selection for this study is located at the Evaluation and Verification Facility (EVF), Blok 60, Complex Dengkil, Malaysian Nuclear Agency as shown in Figure 1. In the initial stage of fieldwork data detection, the EML RD8100 and GPR GSSI equipment are installed and operated. In this stage, some of the adjustments and calibration on both the equipment are configured to achieve the required standard.

The antenna used by GPR GSSI is 400 MHz frequency which is capable of detecting both metallic and non-metallic underground objects. The 400 MHz antenna is considered a medium high frequency which can penetrate from 2.0m to 5.0m depth beneath the ground. GPR has been calibrated by measuring the real depth of the targeted MS pipe to the road surface which is visible from the inspection sump shown in Figure 1 and Figure 2. The measurement was taken and calculated based on the distance of the top of the MS pipe to the top curb, minus the distance of the road surface to the top of curb as shown in Equation (1).

$$Top\ of\ MS\ pipe\ to\ top\ curb - Top\ of\ road\ surface\ to\ top\ curb = Ground\ Truth\ Depth \quad (1)$$

After the first calibration, a GPR scan was done in the study area to locate the MS pipe and taken as the Ground Truth Depth that will be used in equations (2) and (3), followed by an EML scan which collected the Measured Depth used in the same equations.

The research methodology can be simplified by dividing it into three (3) main stages.

Stage 1 : Data collection by using GSSI GPR and EML RD 8100 shown in Figure 3 to determine the depth of the mild steel pipe at the survey Area.

Stage 2 : Data processing in which the data collection by GPR will be post-processed using RADAN software and EML is using Google Earth by recording the GPS data on the site as shown in Figure 4.

Stage 3 : Data interpretation and analysis in terms of accuracy between two (2) types of instruments using statistical error analysis shown in Table 1 to determine the consistency and capability of both utility instruments between GPR and EML.

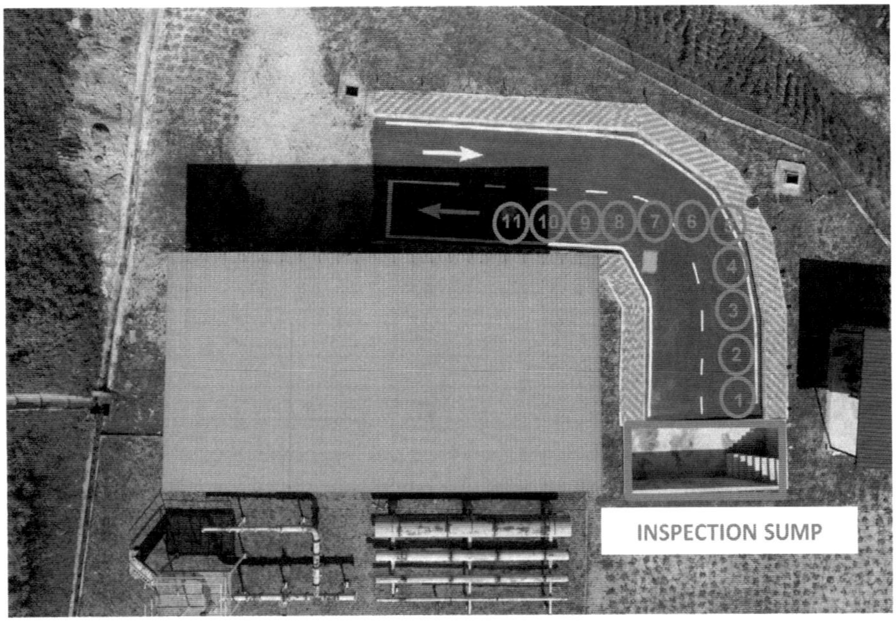

Figure 1. The study area and data collection points located in Blok 60, Malaysian Nuclear Agency.

Figure 2. Distance between the top of the MS Pipe to the top of the curb measured from the inspection sump.

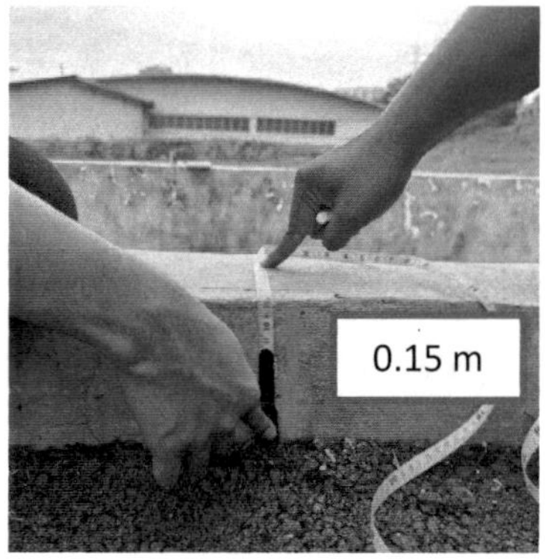

Figure 3. Distance between the curb and the road surface.

Table 1. Error analysis formula use in this study.

Error Analysis		
Root Mean Square Error (RMSE)	$\frac{\sqrt{[\Sigma(Measured\ Depth - Ground\ Truth\ Depth)^2]}}{n}$	(2)
Mean Square Error (MSE)	$\frac{[\Sigma(Measured\ Depth - Ground\ Truth\ Depth)^2]}{n}$	(3)
Absolute Depth Error (ADE)	$EML - GPR = ADE$	(4)

4. Result and discussion

This section serves as the critical juncture where the outcomes of our utility detection and correlation methodology are unveiled and comprehensively analysed. Through the utilization of advanced techniques like GPR and EML, the intricate network of subsurface utilities has been meticulously mapped and assessed. This section delves into the empirical findings and their implications, offering insights into the accuracy, effectiveness, and real-world applications of our methodology. Additionally, it provides a platform for scrutinizing the synergistic advantages of combining GPR and EML data, thus advancing our understanding of how these technologies can be harnessed to optimize utility management, mitigate risks, and enhance the efficiency of construction and excavation projects.

11 points of data have been recorded as shown in Table 2. For data collected using GPR GSSI, the radargrams (hyperbolic images) have been post-processed and analysed with using RADAN 7 software to determine the depth of utility objects. Statistical analysis with error analysis has been used and applied to identify the accuracy of data. Tables 2 shows the result, while Figure 4 shows a graph analysis of data obtained by GPR GSSI and EML RD 8100.

Table 2. Calculated value for ADE, MSE and RMSE based on Point no values.

POINT NO	EML(m)	GPR(m)	ADE	MSE	RMSE
1	1.41	1.15	0.26		
2	1.46	1.15	0.31		
3	1.36	1.15	0.21		
4	1.50	1.20	0.3		
5	1.38	1.20	0.18		
6	1.37	1.15	0.22	0.03727	0.05821
7	1.23	1.15	0.08		
8	1.28	1.15	0.13		
9	1.14	1.15	-0.01		
10	1.23	1.15	0.08		
11	1.46	1.50	-0.04		

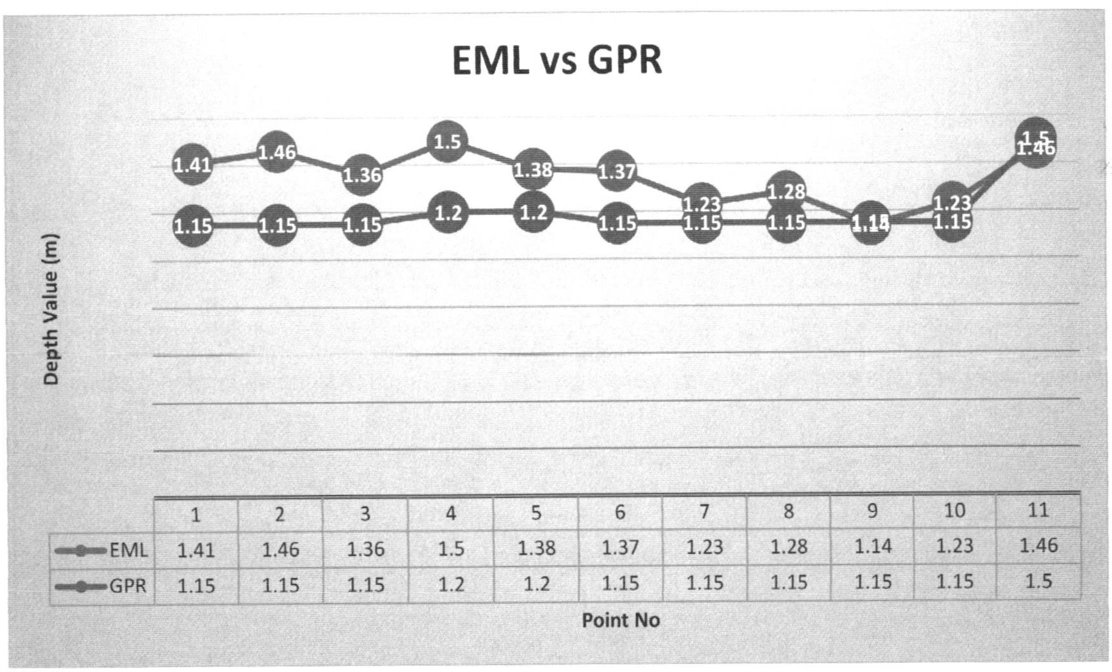

Figure 4. Graph analysis for EML vs GPR based on Depth value.

Based on Figure 4, GPR shows more consistency with minimum variation in the depth readings. EML survey shows accurate results in locating the pipe horizontally but shows some variations in the depth readings. For GPR equipment used, the antenna was shielded perfectly and was free from noise or other disturbances from metallic and non-metallic underground objects. On the other hand, the data obtained from EML RD8100 uses the electromagnetic induction method is less accurate and inconsistent. This is because the electromagnetic signal induced by the EML can easily be disturbed by any metallic objects near the study area, causing it to jump or fade to the non-targeted metallic objects.

Based on the error analysis by comparing GPR and EML depth readings, the average RMSE value is 0.05821. This value shows that the EML reading is acceptable and can complement the GPR reading in terms of utility locating purposes.

5. Conclusion

The site selection of the EVF facility, Blok 60, Malaysian Nuclear Agency is relevant as a study area since the depth and location of underground utilities are known and can be measured for data verification. The study is carried out to determine the accuracy of depth values obtained from GPR GSSI and EML RD8100 which are based on radar and electromagnetic concepts respectively. During the determination of accuracy using GPR GSSI and EML RD8100, it indicates that not all buried pipes, cables and ducts can be detected and mapped in the certain consideration of depth, location, material type, geology, and proximity to other utilities due to limitation and difficulty in determining the actual depth. Based on the result and analysis, it can be concluded that the accuracy achieved from GPR GSSI is almost equal to EML RD8100. It has been proved by statistical data analysis in which the Root Mean Square Error (RMSE) values from 0.05821.

In terms of the reliability of the dataset for both types of equipment for underground utility mapping, it shows that the range value of depth for all the types of utility between GPR GSSI and EML RD8100 is not significantly diverse. In other words, both types of equipment can complement each other's information and improve the ability to detect depth, it is also able to reduce the risk of false positives or false negatives. This is a good sign as a firm decision can be made for good.

6. References

[1] Lopez A, Bacelar R, Pires I, Santos T G, Sousa J P and Quintino L 2018 Non-destructive testing application of radiography and ultrasound for wire and arc additive manufacturing *Addit. Manuf.* **21** doi: 10.1016/j.addma.2018.03.020.

[2] Vestenický P, Mravec T and Vestenický M 2017 Comparison of various marker localization methods *Adv. in Intelligent Sys. and Comp.* 461 doi: 10.1007/978-3-319-44354-6_6.

[3] Young G, Hanson D, Sjostrom K, Jones R and Clark J 2009 Multi-sensor utility mapping in all soils and conditions *SEG Tech. Prog. Expanded Abstr.* **28**(1) doi: 10.1190/1.3255130.

[4] Siu K L and Lai W WL 2013 Principle and type of acoustic emission sensors *Geophys. Sensor Arrays* 1456–1469,.

[5] Conyers L B 1998 Ground penetrating radar exploration and mapping techniques for garden archaeology.

[6] Dusan P and Aleksandar R 2005 Principles of using ground penetrating radar and gps technology for detection of underground utilities.

[7] Tsunasaki M, Hiromori S andNakauchi T 2020 GPR and pipeline & cable locator development *37th Int. NO-DIG Conf. and Exhibit. 2019.*

Acknowledgement

The authors want to express their deep thanks to their colleagues and technical team for their consistent help and encouragement, which have been very important in developing this work. The expertise and support of these individuals have been extremely valuable throughout the research process. Additionally, the authors want to show their sincere gratitude to the Non-Destructive Testing – Material and Structure Integrity Group (NDT-MSI), Agensi Nuklear Malaysia as well as students from Universiti

Malaysia Perlis (UniMAP) for their unwavering support and guidance. Their knowledge, resources, and dedication have significantly contributed to the progress of this project, and their collaborative efforts have played a crucial role in its success. This acknowledgement emphasizes the significance of their partnership and shows how much the authors appreciate their ongoing assistance.

IOP Conf. Series: Materials Science and Engineering 1308 (2024) 012014 doi:10.1088/1757-899X/1308/1/012014

Changes in crystallinity of sago palm (*Metroxylon sago spp.*) waste during cellulose extraction

N. Yacob[1,a)], M.R. Yusof[1], A.Z. Mohamed[2] and K. Badri[3]

[1]Malaysian Nuclear Agency (Nuclear Malaysia) Bangi, 43000 Kajang, Selangor, Malaysia
[2]Programme of Pulp and Paper & Pollution Control, Laboratory of Biopolymer and Derivatives, Institute of Tropical Forestry and Forests Products (INTROP), Universiti Putra Malaysia, 43400 UPM Serdang, Selangor, Malaysia
[3]School of Chemical Sciences, Faculty of Science and Technology, Universiti Kebangsaan Malaysia, 43560 Bangi, Selangor, Malaysia

[a]norzita@nm.gov.my

Abstract Sago palm (*Metroxylon sago spp.*) is one of the major agricultural plants in Malaysia. Sago pith waste (SPW), which is left behind after sago starch extraction, is disposed of by washing off into the nearby river. Reuse of SPW provides a solution to the problem of discarding the waste. In this study, cellulose has been extracted from SPW. The fibers were pretreated with 2% of aqueous sodium hydroxide (NaOH) solution before bleaching with an acidified sodium chlorite ($NaClO_2$) and acetate buffer solution. The influence of NaOH and bleaching treatment on the cellulose crystallinity of SPW has been investigated using chemical analysis, Fourier Transform Infrared Spectroscopy (FTIR) and X-ray Diffraction (XRD). From XRD analysis, the crystallinity index (CI) was calculated. The CI of SPW increased significantly by 110% after the chemical treatments. The FTIR analysis proved the dissolution of the non-cellulosic components by NaOH and $NaClO_2$ treatments.

1. Introduction

Lignocellulosic comprises of cellulose, hemicellulose and lignin. Cellulose is a long network of glucose molecules linked by β $(1 \rightarrow 4)$ glycosidic bonds. It consists of both amorphous and crystalline regions. Hemicellulose is a macromolecule branch with short chains containing various sugars. Thus, hemicellulose has a lower molecular weight compared to cellulose. Lignin is a complex three-dimensional polymer that binds both cellulose and hemicellulose [1-3]. Thus, to extract cellulose from lignocellulosic materials, lignin and hemicellulose have to be separated.

Crystallinity has a significant impact on the chemical, mechanical and physical properties of cellulose. An increase in crystallinity will lead to an increase in mechanical properties such as tensile strength and dimensional stability. However, some properties such as swelling and solubility will decrease. Hence, it is important to know the amount of crystallinity and amorphous phase in the cellulose. According to Terinte et al. [4], the CI measurement depends on the instrument used, the data interpretation technique and the sample quality. Among the instruments used for CI measurements are

X-ray Diffraction (XRD), solid state ^{13}C Nuclear Magnetic Resonance (NMR) and Raman Spectroscopy.

The most effective technique for measuring the CI of the cellulose is XRD. There are three methods commonly used based on XRD data as illustrated in Figure 1 [5]. The peak height method is an empirical technique and is the most common and simplest method for CI measurement developed by Segal et al. [6]. It is based on the height ratio of the highest crystalline peak (002) at 2θ=22° and the height of the amorphous peak at 2θ=18° (Figure 1a). The second method is based on the peak deconvolution (Figure 1b). The ratio of the area of crystalline peaks to the total area of the cellulose is calculated to determine the CI [7]. The third method is the amorphous subtraction method (Figure 1c). The CI is determined from the ratio of the area above the amorphous sample to the total area using the amorphous standard [8].

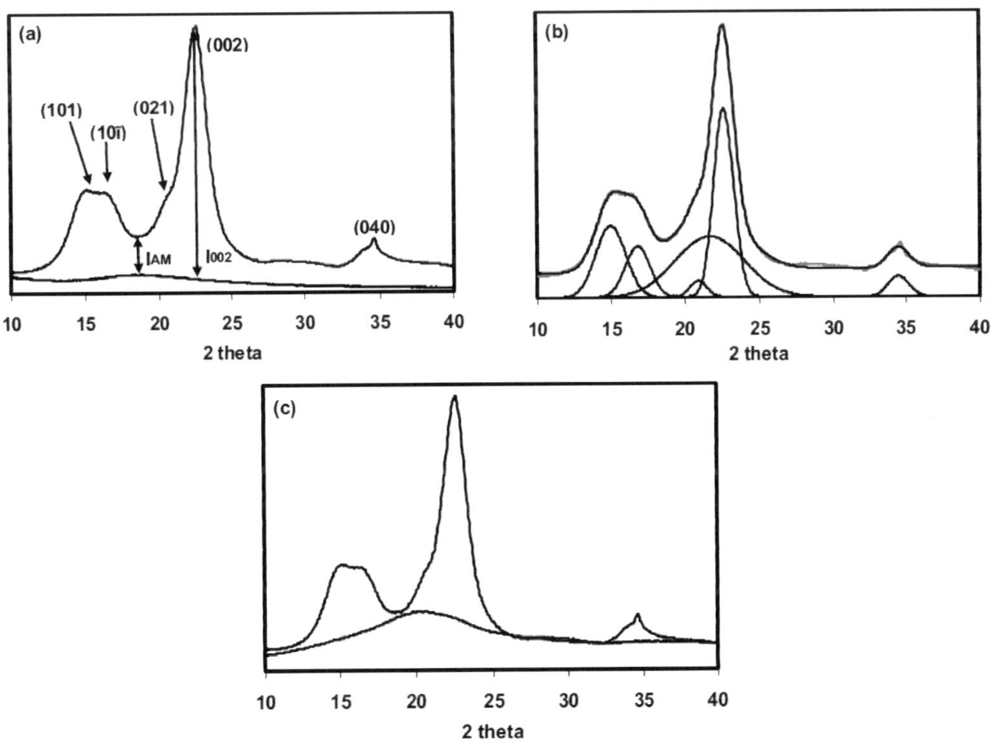

Figure 1. Determination CI using XRD (a) peak height, (b) peak deconvolution and (c) amorphous subtraction method.

In this study, cellulose was extracted from sago palm (Metroxyon sago spp.) pith waste (SPW) by alkaline hydrolysis followed by bleaching using acidified sodium chlorite delignification. The aim was to investigate the effect of the chemical treatment on the crystallinity and crystal structure of the SPW. We studied by determining their chemical composition (TAPPI standard), chemical structure by using Fourier Transform Infrared (FTIR) spectroscopy and crystal structure by X-ray diffraction (XRD).

IOP Conf. Series: Materials Science and Engineering 1308 (2024) 012014 doi:10.1088/1757-899X/1308/1/012014

2. Method

2.1 Materials

SPW was supplied by Ng Kia Heng Sago Industry Bhd. (Batu Pahat, Johor). Sodium chlorite, $NaClO_2$ and acetic acid, CH_3COOH were supplied from Sigma-Aldrich Inc. (St. Louis, MO, USA). Sodium hydroxide NaOH, was purchased from Fisher Scientific Inc. All reagents and solvents were used as received.

2.2 Chemical composition

A chemical composition which is holocellulose, α-cellulose and Klason lignin were estimated for both untreated and treated SPW. Holocellulose was determined using the method developed by Wise et. al [9]. The hemicellulose content was examined by subtracting α-cellulose from the holocellulose part whereas the α-cellulose was determined by TAPPI standard methods T203. An extractive content was calculated according to TAPPI T204 whereas Klason lignin content was estimated according to TAPPI standard method T222. Total ash content was examined by weighing the residue remaining after combustion at 550 °C until all the carbon was eliminated.

2.3 Extraction of cellulose from sago pith waste (SPW)

Cellulose was extracted from SPW as proposed by Yacob et al. [10] with some modifications. First, SPW was stirred at 250 rpm with 2 % NaOH aqueous solution at 60 °C for 1 h. At this stage, alkaline treated SPW was achieved. Subsequently, the alkaline-treated SPW underwent bleaching treatment. The alkaline treated SPW was stirred with a solution prepared with equivalent parts of 1.7 (w/v) aqueous $NaClO_2$ in water, acetate buffer and distilled water at 80 °C for 4 h. At this stage, cellulose had been successfully extracted from SPW. Each treatment was washed with deionized water and filtered until pH 7 before drying at 40° until constant weight. The cellulose was then freeze dried and kept in a desiccator before further testing.

2.4 Determination of Changes in Functional Groups Using Fourier Transform Infrared Spectroscopy (FTIR)

The changes in functional groups of untreated and treated SPW were examined using FTIR analysis. The FTIR spectra were recorded on a spectrometer model Bruker Tensor II, Germany in the range of 4000 to 500 cm^{-1} with a scanning resolution of 2 cm^{-1}.

2.5 Determination of Crystallinity Index (CI) using X-ray Diffraction (XRD)

The effect of chemical treatments on the crystal structure of SPW was investigated by XRD. The X-ray diffractograms of untreated and treated SPW were obtained using an RÖntgen diffractometer system (PANalytical. Netherlands). The diffracted intensity of Cu-Kα radiation was measured at 40 kV, 30 mA at 25 °C. The scattering angle was measured at 2θ from 5° to 40° at a 2 °C/min scanning rate. Gaussian deconvolution was used for peak separations. The peak height method was used to determine the CI calculated using the Segal method given by Equation (1):

$$CI = \frac{I_{002} - I_{am}}{I_{002}} x\ 100 \quad (1)$$

where I_{002} and I_{am} are the peak intensities of crystalline at 2θ=22° and amorphous at 2θ=18° respectively. The crystallite size (L) was calculated using Scherrer Equation (2):

$$L = \frac{K * \lambda}{\beta * \cos\theta} \quad (2)$$

where K is a constant of value 0.94, λ is the X-ray wavelength (0.154 nm), β is the half-height width of the diffraction band and θ is the Bragg angle corresponding to the (002) plane.

3. Results and discussion

3.1. Chemical composition

The chemical compositional analysis data of untreated, alkaline treated and bleached SPW is displayed in Table 1. The major component in SPW consists of starch granules which 38% were entrapped within the fibers during starch extraction followed by hemicellulose (17%), cellulose (12%), lignin (4%) and other constituents. After bleaching, cellulose significantly increased to 500% while hemicellulose decreased to about 55%. No extractive and lignin were detected as well as starch content after bleaching. Hemicellulose was hydrolyzed during alkaline treatment by breaking down the polysaccharides, hence releasing the cellulose fibers [11]. Whereas, lignin was removed upon bleaching by breaking down the glycosidic β-ether bonds into a simpler compound.

Table 1. Compositional analysis data of untreated SPW and treated SPW.

	Moisture content (%)	Starch (%)	Cellulose (%)	Hemicellulose (%)	Lignin (%)	Extractive (%)	Ash (%)
SPW	3.50 ± 0.17	38.41 ± 1.36	14.31 ± 1.48	17.45 ± 0.90	4.83 ± 1.04	2.82 ±1.36	2.74 ± 0.42
Alkaline treated SPW	2.01 ± 0.23	7.05 ± 0.23	63.26 ± 1.50	21.02 ± 3.21	3.03 ± 0.89	-	1.34 ± 0.13
Bleached SPW	1.90 ± 0.90	-	91.26 ± 2.84	7.84 ± 1.32	-	-	-

3.2. FTIR analysis

The removal of non-cellulosic components after the treatment was evidenced by FTIR as illustrated in Figure 2 through the loss of peaks corresponding to hemicellulose and lignin. Figure 2a depicts the untreated SPW. A peak that occurred at 3311 cm^{-1} was associated with the stretching vibration of O-H groups. This peak was observed in all samples. The band at 2900 cm^{-1} was due to the symmetric C-H vibrations. Whereas, the peak approximately at 1711 cm^{-1} was ascribed to the C=O vibration. This peak referred to acetyl and uronic ester groups of hemicellulose. Other than that, this peak relates to the ester linkage of the carboxylic group of the ferulic and *p*-coumaric acids of lignin and/or hemicellulose as well [12]. Peaks at 1532 cm^{-1} and 1236 cm^{-1} as shown in Figure 2a were attributed to C=C and C-O aromatic rings in lignin [13].

The disappearance of peaks at 1711 cm^{-1}, 1532 cm^{-1} and 1236 cm^{-1} was depicted in Figure 2b and Figure 2c. The disappearance was due to the removal of hemicellulose and lignin after alkaline hydrolysis and bleaching. This was supported by the results in Table 1. The treatments dissolved non-cellulosic components by cleaving the α- and β- aryl ether linkages between lignin and hemicellulose.

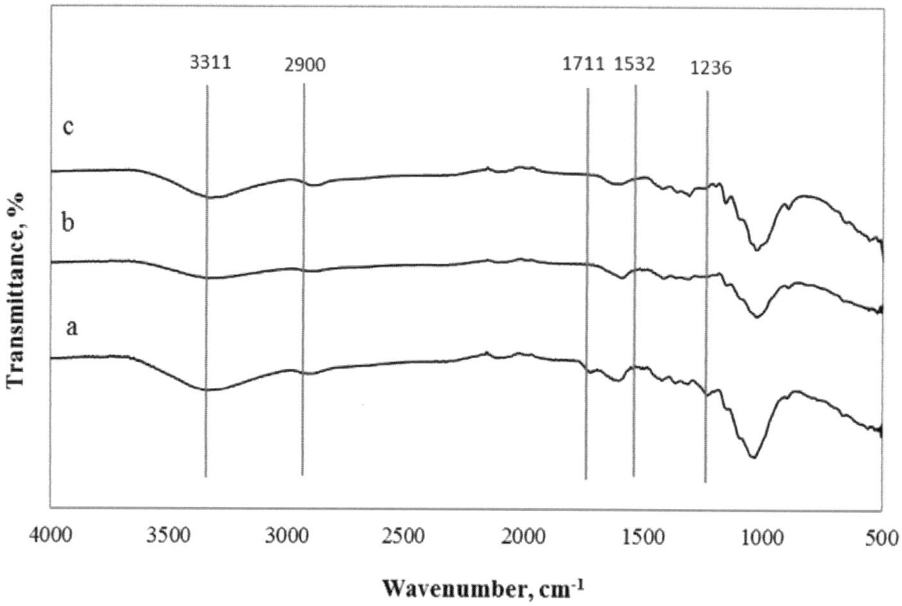

Figure 2. FTIR spectra of (a) untreated, (b) alkaline treated and (c) bleached SPW.

3.3. XRD analysis

Generally, XRD is used to evaluate the degree of crystallinity of wood and fibers. To examine the intensity of the diffraction bands, Gaussian deconvolution was performed. Figure 3 shows the XRD diffractogram of untreated and treated SPW. As shown in Figure 3a, the diffractogram displayed broad peaks 2θ value at 17° and 23° and medium peaks at around 2θ value at 5.3° and 15.0°. Similarly, a few peaks scattered around 2θ values at 10.2° and 11.7. These peaks referred to fibers and sago starch granules that remained in the parenchyma cell of SPW [14-15]. This was in agreement with chemical compositional analysis as shown in Table 1. Figure 3b and Figure 3c revealed peaks 2θ value at 15.6°, 22.7° and 34.6° for the treated SPW. The 2θ = 15.6° was assigned to the (110) crystallographic planes whereas 2θ = 22.7 and 34.6° corresponded to the (002) and (004) planes respectively. The bleached SPW showed a narrower and sharper peak as compared to alkaline treated SPW. These peaks represented the crystalline peak of cellulose I [16]. The XRD diffractograms of alkaline treated and bleached SPW displayed a similar pattern indicating that the crystal structure of cellulose fibers is not changed during the treatment.

On the other hand, the intensity of the diffraction peaks showed a slight difference. The CI and crystallite size of untreated and treated SPW are presented in Table 2. For treated SPW, narrower and more intense peaks were seen. The CI of SPW increased by 110% from 40% to 84% after bleaching. It shows that the amorphous region of the fibers is removed during the alkaline hydrolysis. The removal was more significant after bleaching treatment [17]. The elimination of the amorphous region was in agreement with the results shown in FTIR spectra (Figure 2). On the contrary, the crystallite size of SPW slightly decreased after each treatment because the surface of the crystallite corresponding to amorphous regions weakened during the treatment [18]. These results suggest that bleached SPW has a more compact crystalline structure.

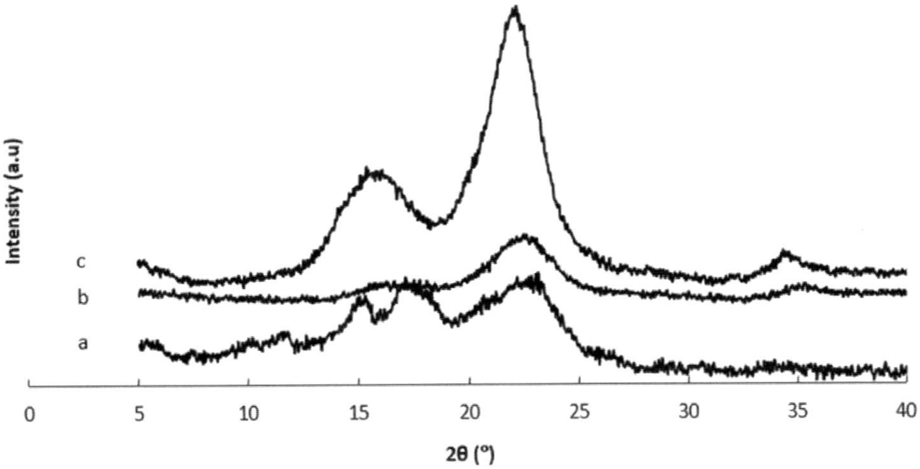

Figure 3. XRD diffractogram of (a) untreated (b) alkaline treated and (c) bleached SPW.

According to Garvey *et al.* [19], wood cellulose is a combination of two different crystalline structures, namely Iα and Iβ. The Iα structure is assigned to triclinic, while Iβ has monoclinic unit cells. The Iβ has a more stable structure than that of Iα. It is represented by several intense peaks at ($1\bar{1}0$), (110), (102), and (004). As a consequence, it is proven that the SPW has a cellulose Iβ structure.

Table 2. CI and crystallite size of SPW, alkaline treated SPW and bleached SPW.

	Crystallinity Index (%)	Crystallite size (nm)
SPW	40	3.49
Alkaline treated SPW	48	3.31
Bleached SPW	84	3.12

4. Conclusion

This study explored the feasibility of applying alkaline hydrolysis followed by bleaching to extract cellulose from SPW. The obtained results indicated that the cellulose was successfully extracted from SPW and the content increased from 14% to 91%. FTIR and XRD analysis confirmed the removal of non-cellulosic components of SPW. Furthermore, crystallinity analysis showed that the crystal structure of the cellulose does not change during the chemical treatments.

5. References

[1] Bajpai P. 2016 Structure of lignocellulosic biomass: Pretreatment of lignocellulosic biomass for biofuel production (Singapore: Springer Publishing).

[2] Ahmazadeh A and Zakaria S 2009 Preparation of novolak resin by liquefaction oil palm empty fruit bunch (EFB) into phenol and characterization of EFB residue *Polimer Plastic Technol. and Eng.* **48** 10-6.

[3] Nechyporchuk O, Belgacem M N and Bras J 2016 Production of cellulose nanofibrils : A review of recent advances *Indust. Crops & Prod.* **93** 22–5.

[4] Terinte N., Ibbett, R. and Schuster, K.C. 2011 Overview on native cellulose and microcrystalline cellulose I structure studied by X-ray diffraction (WAXD): Comparison between measurement techniques *Lenzinger Berichte 89* **1** pp 118-31
[5] Park S, Baker J O, Himmel M E, Parilla P A and Johnson D K 2010 Cellulose crystallinity index: measurement techniques and their impact on interpreting cellulase performance *Biotechnol. for Biofuels* **3**(1) 10.
[6] Segal L, Creely J J, Martin A E J and Conrad, C M 1958 Empirical method for estimating the degree of crystallinity of native cellulose using the X-Ray diffractometer *Textile Res. J.* 29 **10** pp 786–94
[7] Hult L E, Iversen T and Sugiyama J 2003 Characterization of the supramolecular structure of cellulose in wood pulp fibres *Cellulose* **10** 103-10.
[8] Bansal P, Hall M, Realff M J, Lee J H and Bommarius A S 2010 Multivariate statistical analysis of X-ray data from cellulose: A new method to determine degree of crystallinity and predict hydrolysis rates *Bioresource Technol.* **101** 4461-71.
[9] Wise L E 1946 Chlorite holocellulose, its fractionation and bearing on summative wood analysis and on studies on the hemicelluloses *Paper Trade* **122** 35–43.
[10] Yacob N, Yusof M R, Mohamed A Z and Badri K H 2018 Effect of cellulose fiber from sago waste on properties of starch-based films *IOP Conf. Series: Mater. Sci. and Eng.* 368.
[11] Neto W P F, Silvério H A, Dantas N O and Pasquini D 2013 Extraction and characterization of cellulose nanocrystals from agro-industrial residue – Soy hulls *Indust. Crops and Prod.* **42** 480-8.
[12] Melikoğlu A Y, Bilek S E and Cesur S 2019 Optimum alkaline treatment parameters for the extraction of cellulose and production of cellulose nanocrystals from apple pomace *Carbohyd. Polym.* **215** 330-7
[13] Fareez I M, Ibrahim N A, Yaacob W M H W, Razali N A M, Jasni A H, and Aziz F A 2018 Characteristics of cellulose extracted from Josapine pineapple leaf fibre after alkali treatment followed by extensive bleaching *Cellulose* **25**(8) 4407–21.
[14] Ahmad F B, Williams P A, Doublier J L, Durand, S and Buleon A 1999 Physico-chemical characterisation of sago starch *Carbohydr. Polym.* **38(4)** 361–70.
[15] Lai J C, Rahman W A W A and Toh W Y 2014 Mechanical, thermal and water absorption properties of plasticised sago pith waste *Fibers and Polym.* **15**(15) 971–978.
[16] Gopi S, Amalraj A, Jude S, Benson K T, Balakashnan P, Haponiuk J T and Thomas S 2019 Isolation and characterization of stable nanofiber from turmeric spent using chemical treatment by acid hydrolysis and its potential as antimicrobial and antioxidant activities *J. of Macromolecul. Sci. Part A* **56**(4) 327–40.
[17] Tibolla H, Pelissari F M, Rodrigues M I, Menegalli F C 2017 Cellulose nanofibers produced from banana peel by enzymatic treatment: Study of process conditions *Industr. Crops and Prod.* **95** 664–74.
[18] Poletto M, Zattera A J, Forte M M and Santana R M 2012 Thermal decomposition of wood: Influence of wood components and cellulose crystallite size *Bioresour. Technol.* **109** 148-53.
[19] Garvey C J, Parker I H, Simon G P 2005 On the interpretation of X-ray diffraction powder patterns in terms of the nanostructure of cellulose I fibers *Macromol Chem. and Phys.* **206** 1568-75.

Acknowledgement
The authors would like to thank to Radiation Processing Division and Industrial Technology Division, Malaysian Nuclear Agency for technical support and equipment facilities.

IOP Conf. Series: Materials Science and Engineering 1308 (2024) 012015 doi:10.1088/1757-899X/1308/1/012015

Analysis of Thorium (Th) and Uranium (U) in radioactive wastewater sample in accordance with American Public Health Association (APHA) 3125 method

Munirah Abdul Zali[1,a], Azian Hashim[1], Shakirah Abd Shukor[1], Siti Aminah Omar[1], Zalina Laili[1]

[1]Analytical Chemistry Application Group, Waste & Technology Division, Malaysian Nuclear Agency, 43000 Kajang, Selangor

[a]munirahzali@nm.gov.my

Abstract. The main issue with thorium (Th) and uranium (U) in radioactive wastewater is the potential of these elements to contaminate the water bodies. The unregulated radioactive wastewater release can affect the ecosystem and potentially harm aquatic life. Therefore, there is a need to establish the analysis of Th and U using standard method to obtain better sensitivity and good confidence in the final report. In this study, the determination of uranium and thorium was employed using the American Public Health Association (APHA) method 3125 standard method. The method performance and method detection level (MDL) samples for APHA 3125 were prepared in the deionized water and digested using hot block (APHA 3030E) and microwave digester (APHA 3030K). The quantification of ^{238}U, ^{235}U and ^{232}Th in the method performance and MDL samples was performed using Inductively Coupled Plasma-Mass Spectrometry (ICP-MS). From the results, the method performance samples for both digestion methods were in accordance with the APHA requirement. The MDL obtained from APHA 3030E for ^{232}Th, ^{238}U, and ^{235}U were 5, 7, and 8 μg L^{-1} respectively. Meanwhile, the MDL obtained from the microwave digester for ^{232}Th, ^{238}U, and ^{235}U, and were 2, 3, and 3 μg L^{-1} respectively. The APHA 3030K digestion technique gives lower and better detection limit compared to the APHA 3030E technique. Based on APHA 3030K MDL, ^{232}Th was detected in the radioactive wastewater sample while ^{238}U and ^{235}U were less than MDL. Meanwhile, for APHA 3030E, all elements were below MDL. This study will enhance the detection of uranium and thorium in the radioactive wastewater sample towards better environmental monitoring programs at the national and global levels.

1. Introduction

The radioactive Thorium (Th) and Uranium (U) elements are ubiquitously present in the earth's crust with established important applications [1]. Concerns over release of these elements and their possible ecological effect on the environment are increasing due to rapid urbanization and industrialization. Various agencies or committees such as the World Health Organization (WHO) have introduced generic guideline values relating to the maximum levels of these elements that can be released into or be contained in the environment [2]. Nowadays, Th and U are fundamental in critical technologies, thus the uses and extraction of these elements are increasing leading to the discharge into the environment with potential risk to humans and biota [3,4].

Determination of Th and U in water samples such as wastewater, river water, groundwater and tap water can be challenging due to the low-level detection in the majority of the water samples and potential chemical and physical interferences in the samples. In recent years, due to advanced technology of inductively coupled plasma-mass spectrometer (ICP-MS), Th and U quantification was able to be done in low-level samples with fast, reliable and good precision analysis [5,6]. Furthermore, in conjunction with the established standard method such as the American Public Health Association (APHA) method, the Th and U elements quantification can be performed and produce reliable results by following strict method performance criteria and quality control procedures. For Th and U quantification, the APHA 3125 method was the most suitable method for the simultaneous measurement of Th and U in water samples.

Principally in the APHA 3125 method, the water sample is introduced into an argon-based, high-temperature radio frequency plasma by using pneumatic nebulization. The energy from the plasma transfers to the water sample and causes desolvation, atomization, and ionization of elements. Ions generated by these energy-transfer processes are extracted from the plasma by the vacuum interface, and separated based on mass-to-charge ratio by the mass spectrometer [7]. The ions passing through the mass spectrometer are counted by an electron multiplier and the resulting information is processed by a computer-based data-handling system. In APHA 3125, water samples are recommended to be digested using the APHA 3030E method (hot-block digestion). However, this method also allows other digestion techniques such as APHA 3030K by using the microwave digester technique. Furthermore, the two digestion methods based on APHA 3030E and APHA 3030K were required to perform the method performance to evaluate the method capability and verification of the obtained result.

Among the method performance criteria recommended in the APHA method, every batch of analysis requires the determination of method blank (MB), laboratory fortified blank (LFB), laboratory fortified matrix (LFM), laboratory fortified matrix duplicate (LFMD) and relative percentage difference (RPD) between LFM and LFMD in every batch sample in order produce a reliable result [8]. Furthermore, APHA also requires the determination of method detection level (MDL) in every method to set acceptable levels for reporting limit for each Th and U in real environmental samples. This study aims to determine and compare the method performance and MDL for Th and U in water samples using APHA 3030E and APHA 3030K digestion techniques and quantification by using APHA 3125 method. The findings of this study can be applied in the reporting of Th and U in water samples and can be used to implement further action in environmental management pollution protection improvement.

2. Materials and methodology

2.1. Materials

In this study, the standard calibration was purchased from Perkin Elmer, USA, and consisted of 10 µg ml^{-1} of Th and U elements. A single standard of rhodium (Rh) was obtained from the Scharlau brand for internal standard purposes during analysis. Concentrated nitric acid with trace metal grade was purchased from Fisher Chemical. Deionized water (DIW) for washing, soaking and sample preparation was based on the high-quality reagent water specification based on the APHA 1080 method [9].

2.2. Preparation of quality control samples

In this study, two metal digestion methods were used to compare the quality control (QC) samples consisting of method performance and MDL samples for Th and U determination in water samples. The first method is the APHA 3030E method (hot block digestion) [10]. For method performance of APHA 3030E, two blank, laboratory fortified blank (LFB), unspike, laboratory fortified matrix (LFM) and laboratory fortified matrix duplicate (LFMD) samples were prepared during the analysis. The seven replicates of MDL samples for the APHA 3030E method were also prepared and analyzed together with method performance samples. Briefly, for the APHA 3030E method, 10 ml of quality QC samples were added to the 50 ml tube. The samples were added with 0.5 ml of concentrated nitric acid and placed in the block heater. The temperature of the block heater was adjusted to 105°C. The tube caps for all the samples were placed on top of the tube without screwing. The samples were digested for a minimum of 2 hours and more concentrated nitric acid was added until observation of clear solution. Upon complete digestion, the sample volume was adjusted and recorded. The details on the spiking regime for QC samples for method performance and MDL determination based on APHA 3030E were tabulated in Table 1.

Table 1. The spiking regime of QC samples for APHA 3030E method.

QC samples	Method Performance					MDL (Seven replicates)
	Blank	**LFB**	**Unspike sample**	**LFM**	**LFMD**	
Concentration (μgL^{-1})	-	25	-	25	25	25
Volume of sample (mL)	10	10	10	10	45	45
Volume of Concentrated Nitric Acid (mL)	0.5	0.5	0.5	0.5	0.5	0.5
10 μg ml^{-1} Th & U mixture standard (mL)	-	0.025	-	0.025	0.025	0.025

The second method for MDL determination is APHA 3030K (microwave digester) [11]. For method performance of APHA 3030K, two blank, laboratory fortified blank (LFB), unspike, laboratory fortified matrix (LFM) and laboratory fortified matrix duplicate (LFMD) samples were prepared during the analysis. The seven replicates of MDL samples for the APHA 3030E method were also prepared and analyzed together with method performance samples. Briefly 45 ml of quality control samples were added into microwave digester teflon vessels. Then the samples were added with 5 ml of concentrated nitric acid. The vessels were capped, tightened with vessel holders and placed in microwave digester turntable. The digestion procedure was programmed in the microwave digester system. The samples were heated to 160 ± 4 for 25 minutes for the first stage of digestion. For the second stage, the samples were digested with a slow rise to 170 ± 4 in 10 minutes. The vessels were allowed to cool and the solution was transferred into a clean polypropylene tube. The final sample volume was adjusted and recorded. The details on the spiking regime for QC samples for method performance and MDL determination based on APHA 3030K were tabulated in Table 2.

The determination of Th and U in the QC samples was carried out using Inductively Coupled Plasma Mass Spectrometer (ICP-MS) Nexion 350X, Perkin Elmer (USA) at mass 232 for Th (^{232}Th), 238 and 235 for U (^{238}U and ^{235}U) [7]. Briefly, the water samples are introduced to argon-based, high temperature radio-frequency plasma via pneumatic nebulization. As energy is transferred from the plasma to the sample stream, the target elements are dissolved, atomized, and ionized. The resulting ions are extracted from the plasma through a differential vacuum interface and separated based on their mass-to-charge (m/z) ratio via mass spectrometry. The collision cell technology with helium gas is used to improve sensitivity. For ^{232}Th, ^{238}U and ^{235}U quantification, the standard gas mode was set in the ICP-MS software configuration. In the ICP-MS, an electron multiplier detector is used to count the separated ions, and a computer-based data-management system is used to process the resulting information. The internal standard calibration was used for the quantification of ^{232}Th, ^{238}U and ^{235}U elements. Two calibration points of calibration blank and 200 μg l^{-1} were established with correlation of coefficient more than 0.995 [7]. The calibration verification standard with concentration at 100 μg l^{-1} was analyzed for every ten samples and the values ranged between 90 and 110%.

Table 2. The spiking regime of QC samples for APHA 3030K method.

QC samples	Method Performance					MDL (Seven replicates)
	Blank	**LFB**	**Unspike sample**	**LFM**	**LFMD**	
Concentration (μgL^{-1})	-	25	-	25	25	25
Volume of sample (mL)	45	45	45	45	45	45
Volume of Concentrated Nitric Acid (mL)	5	5	5	5	5	5
10 $\mu g\ ml^{-1}$ Th & U mixture standard (mL)	-	0.025	-	0.025	0.025	0.025

2.3. Determination of method performance criteria

In the method performance criteria, blank, LFB, unspike, LFM and LFMD samples were analyzed to check the accuracy of the digestion procedure and ICP-MS measurement. The RPD was determined based on LFM and LFMD concentration to determine the difference between the two duplicate sample results. The calculation of LFB, LFM, LFMD and RPD recovery were as follows:

$$LFB\ \%\ Recovery = \frac{found\ value - blank}{true\ value} \times 100 \tag{1}$$

$$LFM\ \%\ Recovery = \left[\frac{LFM\ concentration \times (spike\ volume + sample\ volume) - (unsspike\ sample\ concentration\ x\ sample\ volume)}{spike\ solution\ concentration\ x\ spike\ volume}\right] \times 100 \tag{2}$$

$$LFMD\ \%\ Recovery = \left[\frac{LFMD\ concentration \times (spike\ volume + sample\ volume) - (unsspike\ sample\ concentration\ x\ sample\ volume)}{spike\ solution\ concentration\ x\ spike\ volume}\right] \times 100 \tag{3}$$

$$RPD\ (\%) = \left[\frac{|LFM - LFMD|}{\left(\frac{LFM+LFMD}{2}\right)}\right] x\ 100 \tag{4}$$

2.4. Determination of method detection level (MDL)

The MDL was prepared according to Table 1 and Table 2 for APHA 3030E and APHA 3030K respectively. The MDL was prepared and analyzed at least seven replicates of this solution over 3 days to ensure that the MDL determination is more representative of routine measurement. The standard deviation (SD) of the seven replicates was calculated and MDL was determined by using the SD and student *t* value by using the formula below:

$$SD = \sqrt{\frac{\sum(x_i - \mu)^2}{N}} \tag{5}$$

$$MDL = 3.143 \times SD \tag{6}$$

where x is the value of data distribution, μ is the population mean, and N is the total number of observations.

2.5. Determination of Th and U in radioactive wastewater sample

The radioactive wastewater sample was collected from the Low-Level Effluent Treatment Plant (LLETP) facility in the Waste Technology Centre (WASTEC), Malaysian Nuclear Agency, Malaysia. The radioactive wastewater sample was collected in the pre-cleaned polypropylene tube and preserved with nitric acid: DIW (1:1) solution until pH fell below 2. The samples were stored in the refrigerator at 4 ± 2°C in the laboratory prior to analysis. For the determination of ^{232}Th, ^{238}U and ^{235}U, the radioactive wastewater sample was tested by spiking mixture of the standard solution with the final concentration of 25 µg l^{-1}. The radioactive wastewater sample was digested using APHA 3030E and APHA 3030K methods as described previously in the determination of the MDL part. The level of ^{232}Th, ^{238}U and ^{235}U in the digested radioactive wastewater sample was analyzed using ICP-MS. The digestion procedure for both methods was evaluated by spike recovery of the standard Th and U solution into radioactive wastewater sample with the formula as below.

$$Spike\ Recovery\ (\%) = \frac{Concentration\ of\ Element - Concentration\ of\ Method\ blank}{Spiking\ COncentration} \times 100 \quad (7)$$

3. Results and discussion

3.1. Method performance of Th and U in water sample

In this study, the method performance was studied to verify the established APHA method in our laboratory. This project involved the determination of ^{232}Th, ^{238}U and ^{235}U in water using two digestion techniques (APHA 3030E and APHA 3030K) and quantitation was done using ICP-MS (APHA 3125). Based on the APHA 3030E digestion method, the range of LFB percentage recovery for ^{232}Th, ^{238}U and ^{235}U were 100.8%, 102.3%, and 100.3% respectively (Table 3). The LFM recovery (%) for of ^{232}Th, ^{238}U and ^{235}U based on the APHA 3030E digestion method were 112.5%, 105.5%, and 103.9% respectively. Meanwhile, in APHA 3030K, the LFB percentage recovery for ^{232}Th, ^{238}U and ^{235}U were 105.6%, 103.8%, and 105.9% respectively (Table 4). The LFM percentage recovery for APHA 3030K digestion method for ^{232}Th, ^{238}U and ^{235}U were 104.1%, 102.9%, and 105.4% respectively. The RPD for LFM and LFMD for all elements were less than 20% for both digestion methods (Table 3 & 4). Based on method performance requirements, the LFB percentage recovery (%), LFM percentage recovery (%) and RPD for LFM and LFM (%) for both APHA 3030E and APHA 3030K fell within APHA requirements, produce good accuracy, good digestion capability and minimum loss of analyte during digestion [8].

IOP Conf. Series: Materials Science and Engineering 1308 (2024) 012015 doi:10.1088/1757-899X/1308/1/012015

Table 3. Method performance result for APHA 3030E digestion technique.

Elements	LFB (%)	APHA requirement range (%) for LFB	LFM (%)	APHA requirement range (%) for LFM	Relative Percentage Difference (%) for LFM and LFMD	APHA requirement for RPD (%)
^{232}Th	110.8	85-115	112.5	70-130	5.1	<20
^{238}U	102.3	85-115	105.5	70-130	2.1	<20
^{235}U	100.3	85-115	103.9	70-130	0.6	<20

Table 4. Method performance result for APHA 3030K digestion technique.

Elements	LFB (%)	APHA requirement range (%) for LFB	LFM (%)	APHA requirement range (%) for LFM	Relative Percentage Difference (%) for LFM and LFMD	APHA requirement for RPD (%)
^{232}Th	105.6	85-115	104.1	70-130	5.3	<20
^{238}U	103.8	85-115	102.9	70-130	2.1	<20
^{235}U	105.9	85-115	105.4	70-130	0.6	<20

3.2. Method detection level of Th and U in water sample

Figure 1 shows the MDL for ^{232}Th, ^{238}U, and ^{235}U in water samples digested according to APHA 3030E and APHA 3030K methods. The MDL obtained from APHA 3030E for ^{232}Th, ^{238}U, and ^{235}U were 5, 7, and 8 μg L^{-1} respectively. Meanwhile, the MDL obtained from the microwave digester for ^{232}Th, ^{238}U, and ^{235}U, and were 2, 3, and 3 μg L^{-1} respectively. Based on the MDL result, the MDL of APHA 3030K for Th and U elements were lower and better than the MDL of APHA 3030E method. The microwave digestion technique was able to achieve better stability and sensitivity at lower concentrations than the hot block digestion technique.

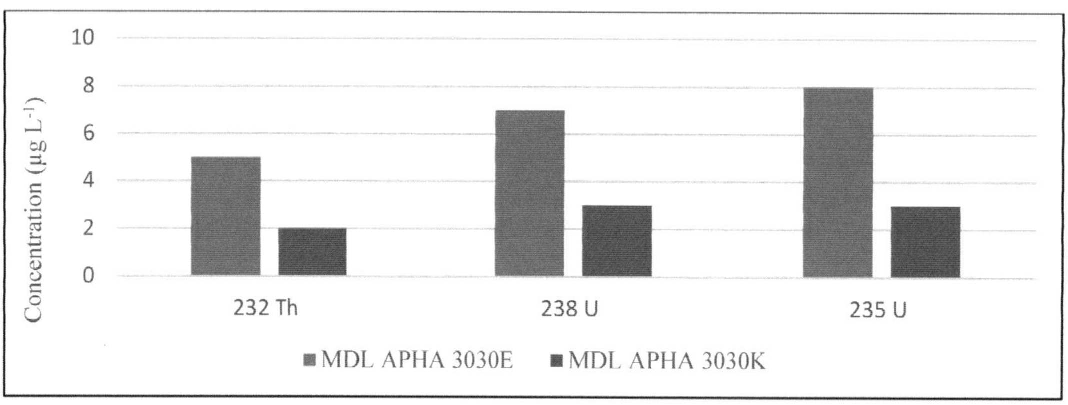

Figure 1. Comparison of MDL of ^{232}Th, ^{238}U, and ^{235}U in water sample.

3.3. Level of Th and U in radioactive wastewater sample

The capability of the APHA 3030E and APHA 3030K digestion methods was tested using real radioactive wastewater sample. Based on Figure 2, the percentage recovery of the spike ^{232}Th, ^{238}U, and ^{235}U in the radioactive wastewater sample was in the range of 70% and 130% for both APHA 3030E and APHA 3030K methods [8]. Both methods show good performance criteria within the APHA requirement for spiking percentage recovery. Furthermore, the percentage recovery value of ^{232}Th, ^{238}U, and ^{235}U in APHA 3030K was closer to a good level compared to the APHA 3030E method. Therefore, the digestion of ^{232}Th, ^{238}U, and ^{235}U in radioactive wastewater sample was better using APHA 3030K than the APHA 3030E method.

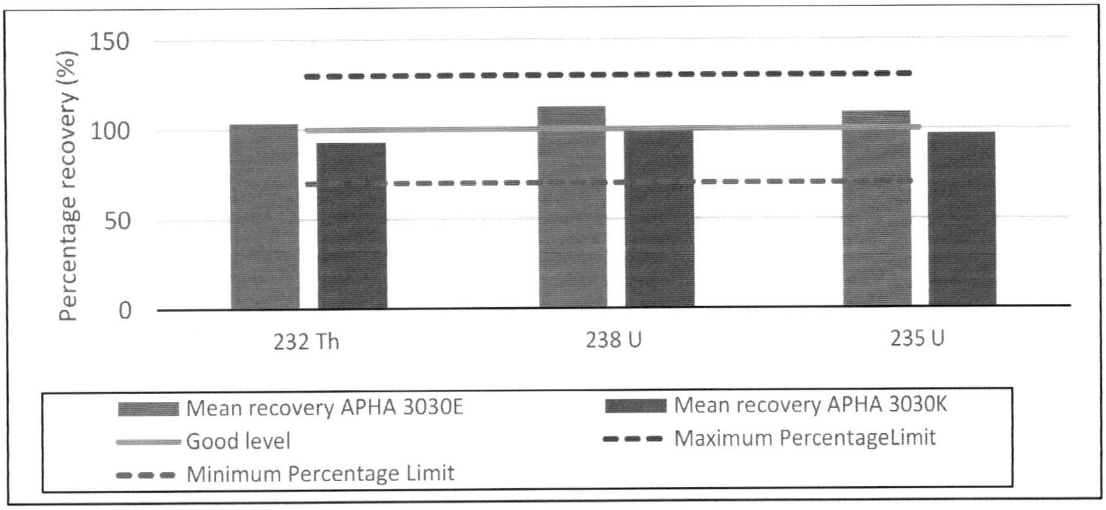

Figure2. Percentage recovery of spike radioactive wastewater sample based on APHA 3030E and APHA 3030K methods.

The determination of the reporting concentration of ^{232}Th, ^{238}U, and ^{235}U in the radioactive wastewater sample must be compared using the MDL of ^{232}Th, ^{238}U, and ^{235}U in water samples with respective digestion methods. From Figure 3(a), the concentration of ^{232}Th, ^{238}U, and ^{235}U in the radioactive wastewater sample based on the APHA 3030E method was lower than the MDL of the APHA 3030E method. Therefore, based on the APHA 3030E method, ^{232}Th, ^{238}U, and ^{235}U were reported less than their respective MDL. In the meantime. In Figure 3(b), ^{232}Th was detected above the MDL value for the APHA 3030K method. The concentration of ^{232}Th in the sample was reported at 3.6 μg L^{-1}. Meanwhile, for ^{238}U and ^{235}U based on the APHA 3030K method, both elements were reported lower than their respective MDL. This finding shows the good capability and sensitivity of the APHA 3030K method compared to the APHA 3030E method. The lower MDL values for ^{232}Th, ^{238}U, and ^{235}U by using the APHA 3030K method than the APHA 3030E method were due to the closed system digestion of the microwave digester compared to open digestion by using hot block digester. The closed system of microwave provides minimum volatilization and stability of elements of interest during the digestion process [12].

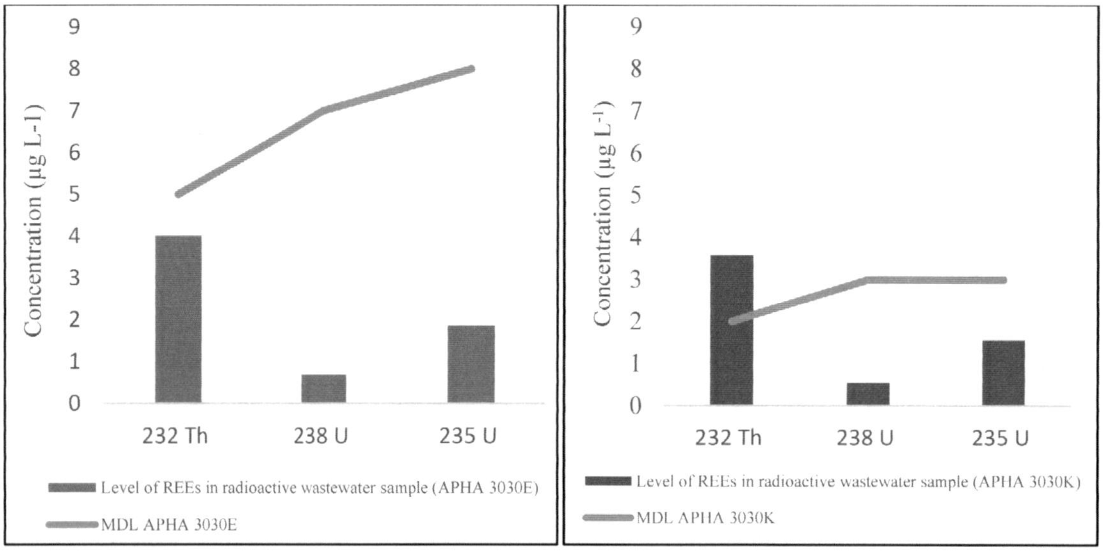

(a) (b)

Figure 3. Comparison of MDL and obtained radioactive wastewater concentration value (a) APHA 3030E; (b) APHA 3030K

4. Conclusions

In this study, two methods of digestion (APHA 3030E and APHA 3030K) for ^{232}Th, ^{238}U, and ^{235}U determination using the APHA 3125 method were performed to determine method performance and MDL for each element in water samples. Based on method performance criteria, both methods produce good value and percentage recovery for LFB, LFM, LFMD and RPD. For ^{232}Th, ^{238}U, and ^{235}U MDL determination, APHA 3030K offers lower value of MDL compared to APHA 3030E. Due to the difference in digestion technique, APHA 3030K offers better sensitivity in the detection of ^{232}Th compared to the APHA 3030E method. Therefore, it is recommended to evaluate method performance criteria and MDL in every digestion method prior to elements quantification to provide desirable reporting levels in various types of water samples.

5. References

[1] Shahin M S A, Saad E A, Shazly A, Ezzat A and El-mongy S A 2020 Improved treatment approaches for trace analysis of Uranium and Thorium using Ultra-sensitive ICP-MS technique *J. Nucl. Radiat. Phys.* **15** 23–33.

[2] Kasar S, Murugan R, Arae H, Aono T and Sahoo S K 2020 A microwave digestion technique for the analysis of rare earth elements, Thorium and Uranium in geochemical certified reference materials and soils by inductively coupled plasma mass spectrometry *molecules* **25.**

[3] Ma L, Dang D H, Wang W, Evans R D and Wang W X 2019 Rare earth elements in the Pearl River Delta of China: Potential impacts of the REE industry on water, suspended particles and oysters *Environ. Pollut.* **244** 190–201.

[4] Soromotin A, Moskovchenko D, Khoroshavin V, Prikhodko N, Puzanov A, Kirillov V, Koveshnikov M, Krylova E, Krasnenko A and Pechkin A 2022 Major, trace and rare earth element distribution in water, suspended particulate matter and stream sediments of the Ob River Mouth *Water* (Switzerland) **14.**

[5] Verplanck P L, Antweiler R C, Nordstrom D K and Taylor H E 2001 Standard reference water samples for rare earth element determinations *Appl. Geochem.* **16** 231–44.

[6] Wysocka I 2021 Determination of rare earth elements concentrations in natural waters – A review

of ICP-MS measurement approaches *Talanta* **221** 121636.
[7] APHA 2022 Metals by inductively coupled plasma *Stand. Meth. for the Examination of Water and Wastewater* 24thed Baxter W (USA: American Public Health Association) 232.
[8] APHA 2022 Quality control practices *Stand. Meth. for the Examination of Water and Wastewater* 24thed Baxter W (USA: American Public Health Association) 190.
[9] APHA 2022 Reagent water *Stand. Meth. for the Examination of Water and Wastewater* 24thed Baxter W (USA: American Public Health Association) 52.
[10] APHA 2022 Nitric acid digestion *Stand. Meth. for the Examination of Water and Wastewater* 24thed Baxter W (USA: American Public Health Association) 196.
[11] APHA 2022 Microwave-assisted digestion *Stand. Meth. for the Examination of Water and Wastewater* 24th ed Baxter W (USA: American Public Health Association) 198.
[12] Alsehli B R M 2021 Evaluation and comparison between a conventional acid digestion method and a microwave digestion system for heavy metals determination in mentha samples by ICP-MS *Egypt. J. Chem.* **64** 869–81.

Acknowledgement
The authors would like thanks to the Analytical Chemistry Application (ACA) Group for the support in budgetary from Trust Account and technical works. The author also would like to thank to WASTEC group for their support during radioactive wastewater sample collection.

IOP Conf. Series: Materials Science and Engineering 1308 (2024) 012016 doi:10.1088/1757-899X/1308/1/012016

Characterisation of sealed radioactive source for nuclear forensic purposes

Zalina Laili[1,a)], Hishamudin Husin[1], Ahmad Hasnulhadi Che Kamaruddin[1], Muhamad Azfar Azman[1], Wilfred Sylvester Paulus[1] and Nadira Kamarudin[1]

[1]Malaysian Nuclear Agency (Nuclear Malaysia), Bangi, 43000 Kajang, Selangor Darul Ehsan, Malaysia

[a]liena@nm.gov.my

Abstract. A study characterized sealed radioactive sources (^{60}Co, ^{137}Cs, and ^{241}Am) for nuclear forensic analysis, focusing on their physical, macroscopic, and radiological properties. Results revealed specific features, such as stamped or engraved information, that served as valuable signatures for nuclear forensic investigation. Corrosion observations provide valuable insights into the condition of the source and historical context. Gamma-ray spectra analysis revealed clear distinctions, including a scattering peak in cylindrical ^{60}Co and ^{137}Cs disk sources (200-500 keV), which was absent in the cup-style ^{241}Am source. Unusual peaks in the gamma spectra of ^{60}Co, ^{137}Cs, and ^{241}Am sources suggested potential impurities or background radiation. This was indicated by non-Gaussian shapes and significant counts per second at specific energy levels. X-ray radiography effectively highlights nuclear material characteristics, emphasizing density, uniformity, and internal structure. This analysis contributed to the development of a National Nuclear Forensic Library (NNFL) and enhanced Malaysia's capability in nuclear forensic investigations related to radioactive sources found outside regulatory control.

1. Introduction

Nuclear forensic analysis aims to determine the origin and history of nuclear and radioactive materials. This analysis involves characterizing samples of nuclear or radioactive material to determine their inherent signature. Validated signatures measured on known material of the same type as the material under investigation assist in interpreting the analytical data obtained in nuclear forensic investigations [1]. This scientific finding can be utilized to answer forensic questions about the materials, such as their composition, origin, production methods, and intended use. In general, the characterization of nuclear and radioactive materials involves determining their specific physical properties, chemical and elemental composition, and isotope ratios, which can be achieved through appropriate analyses. [2].

For sealed radioactive sources, the signatures can provide valuable information for nuclear forensic analysis. Radiological forensic signatures for sealed radioactive sources focus on the physical construction and chemical composition of the radiation capsule and container, as well as physical markings indicating an owner or manufacturer, and isotopically determined material age [3]. Therefore, all characteristics of sealed radioactive sources serve as signatures that can help answer nuclear forensic questions and provide valuable clues to law enforcement agencies about the origin of these sealed radioactive sources that are out of regulatory control.

Sealed radioactive sources are used in various sectors, including industry, medicine, agriculture, and research. Standardized sealed radioactive sources do not exist, and many of these sources are made up of small stainless welded metal components containing radioactive material [3]. Furthermore, significant variations exist across all observable and measurable parameters that characterize these sources. In industrial settings, sealed radioactive sources serve various purposes, including industrial measurements, smoke detectors, irradiation, non-destructive testing, and gamma radiography. The radioisotopes used differ not only in their levels of radioactivity but also in size and mass. Moreover, the construction, materials, and encapsulation methods vary among different types of radiation sources. The physical characteristics of sealed radioactive sources include size, shape, number of capsules, manufacturer, dimensions, weld details, spacer placement, engravings, and construction materials [4]. Consequently, the variations in the physical characteristics of sealed radioactive sources can serve as signatures for specific source models and manufacturers, and can be utilized in nuclear forensic investigations for the source that found outside regulatory control.

Various techniques can be used to identify the signatures of sealed radioactive sources. The selection of a method for characterizing a sealed radioactive source depends on the specific parameters to be determined, whether they are physical, chemical, or radiological. In a nuclear forensics study conducted by Gregor et al., a high-purity germanium detector was used to analyze the gamma-ray spectra of nine different ^{252}Cf sources [5]. In another study, Cummings et al. analyzed a ^{137}Cs source using ICP-MS to determine its age [6]. Notably, nuclear forensic studies are still pending in Malaysia. The present study aims to characterize sealed radioactive sources, specifically ^{60}Co, ^{137}Cs, and ^{241}Am, to identify their inherent signatures. The sealed radioactive sources were characterized according to their physical, macroscopic, and radiological properties. It is significant that no previous studies have been conducted in Malaysia to characterize sealed radioactive sources. Signatures obtained from nuclear forensic analysis are compiled in the Nuclear Forensic Library (NFL). This specialized library is used to identify nuclear and other radioactive materials found outside regulatory control [7]. Therefore, the findings of this study provide valuable insights for nuclear forensic analysis in Malaysia, especially in the establishment of a National Nuclear Forensic Library (NNFL). As a result, this information could help identify or at least narrow down possibilities when dealing with an unregulated radioactive source.

2. Material and methods

2.1. Source description

Four different types of sealed radioactive sources were selected for this study (Fig. 1). Disused Sealed Radioactive Sources (DSRS) were collected from the Waste Technology Development Centre (WasTeC) of the Malaysian Nuclear Agency. The sources include two units of ^{60}Co sources (Source 1 and Source 2), a unit of ^{137}Cs source (Source 3), and a unit of ^{241}Am source (Source 4).

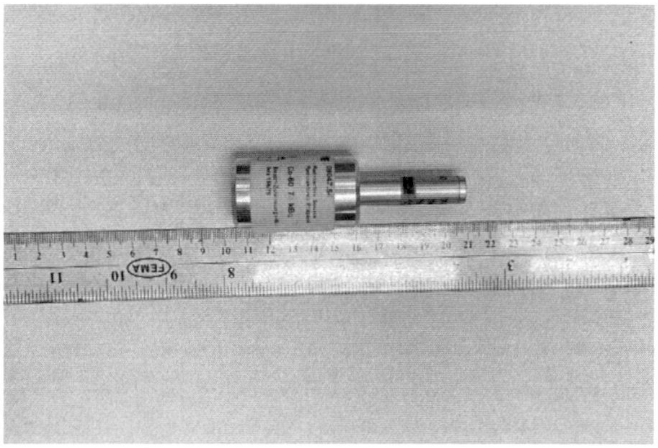

(a) Source 1

(b) Source 2

(c) Source 3

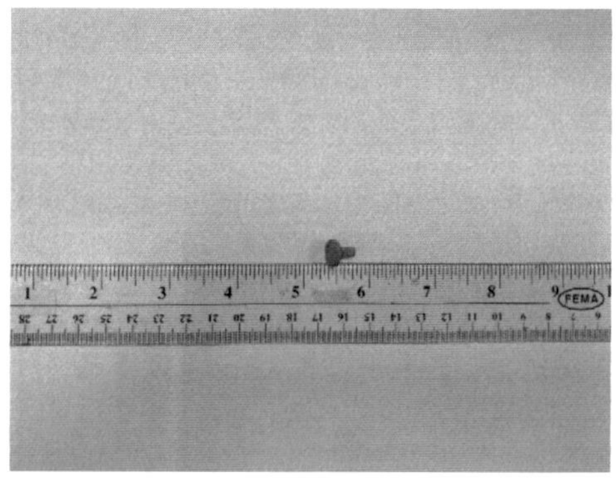

(d) Source 4

Figure 1. Types of sealed radioactive sources used in this study (a) and (b) are ^{60}Co, (c) ^{137}Cs and (d)^{241}Am

2.2. *Characterisation methods*

2.2.1. *Physical characterization*

Physical characterization methods for the source include visual inspection, photography, weight determination, and dimensioning [2]. Visual inspection was conducted to identify the key features visible to the naked eye and to assess the condition of the source. This information includes labeling, engraving, manufacturer, source geometry, and container type. All information gathered during the visual inspection was documented. During the visual inspection, photographs of each source were also taken using a digital camera (Nikon D600). The weight of each source was determined using an analytical balance (Mettler Toledo model AB204-S). The dimensions of the sources were determined using a digital micrometer (Mitutoyo model DC-1MXT) and a digital caliper (Mitutoyo model CD-15APX.

2.2.2. *Gamma spectrometric analysis*

All the sources were measured using a high-resolution coaxial high-purity germanium (HPGe) detector coupled with a multi-channel analyser, featuring a relative efficiency of 25% and a resolution of 1.9 keV at 1332.5 keV of ^{60}Co. The detector model used was EG&G, Ortec, Oak Ridge, TN, USA. Calibration involved a multinuclides standard source, including a range of low to high-energy gamma-rays, such as

IOP Conf. Series: Materials Science and Engineering 1308 (2024) 012016 doi:10.1088/1757-899X/1308/1/012016

^{210}Pb (47 keV), ^{241}Am (60 keV), ^{109}Cd (88 keV), ^{57}Co (122 keV), ^{159m}Te (159 keV), ^{51}Cr (320 keV), ^{113}Sn (392 keV), ^{85}Sr (514 keV), ^{137}Cs (662 keV), ^{60}Co (1173.2,1332.5 keV), and ^{88}Y (898.0 keV,1836.1 keV) [8]. Before starting source counting, three known standard calibration sources with specified activities (^{241}Am, ^{137}Cs, and ^{60}Co) provided by the manufacturer were included for validation and quality control. The peak energy of each calibration source determined the identity of the source. Specifically, ^{241}Am appeared at 60 keV, ^{137Cs} at 662 keV, and ^{60}Co at two energy peaks simultaneously at 1173.2 and 1332.5 keV. Measurements for each source were conducted over an approximate duration of 24 hours, ensuring continuous and reliable evaluation. The acquired spectra were processed and analyzed using specialized spectroscopy software, namely Gamma Vision. The software enables precise measurement of the full width at half maximum (FWHM) of the observed gamma-ray peaks. A smaller FWHM indicates higher resolution, thus improving the discrimination of energy levels. This comprehensive approach enabled the assessment of resolution for each gamma source, providing valuable insight into the precision of energy level discrimination in nuclear forensic investigations.

2.2.3. Radiographic method

Radiographic experiments were performed on sources using an X-ray machine (Dandong Aolong model XXG-2205, China), a directional X-ray tube with a maximum of 200 kV and a fixed current of 5 mA. The film radiography method was selected for this experiment, and digitization and image enhancement were conducted for the purpose of record-keeping. Proper settings were necessary to obtain high-quality images. Thus, the voltage applied ranged from 80 kV to 120 kV, and the exposure time varied from 0.2 up to 0.8 minutes, depending on the source types. Figure 2 shows the setup of the system and the arrangement of objects for radiographic imaging.

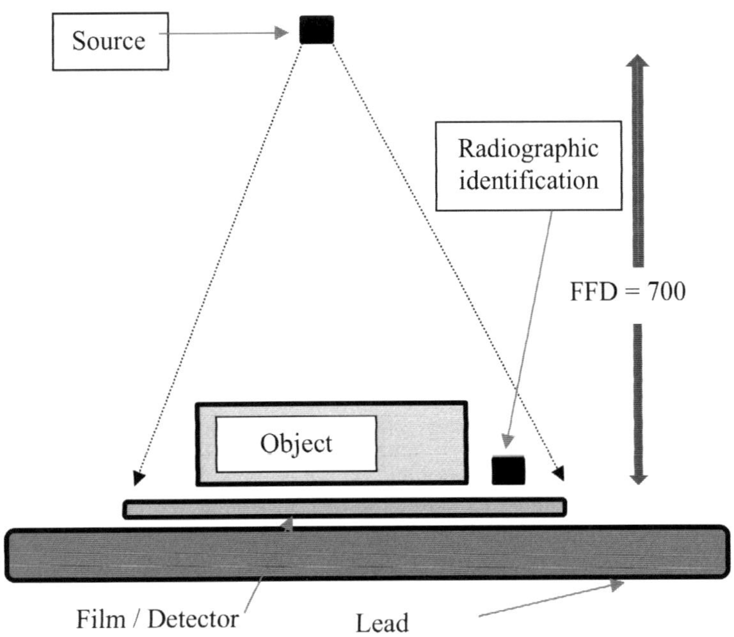

FFD =Focal to film distance

Figure 2. The setup of the system and the arrangement of objects for radiographic imaging.

IOP Conf. Series: Materials Science and Engineering 1308 (2024) 012016 doi:10.1088/1757-899X/1308/1/012016

3. Results and discussion

3.1. Physical characteristic of sources

The physical characterization results are presented in Table 1. Upon visual inspection, notable differences were observed in the construction of each radioactive source. For instance, variations were found in the type of shielding used for the sources. Some sources were safeguarded within sealed stainless-steel capsules, while others consisted of metal foils or discs. Source 1, containing ^{60}Co with an activity of 74 kBq, was in the form of a rod-shaped radioactive source enclosed within a sealed stainless-steel capsule, further held in a metal holder. The rod-shaped source had a cylindrical form (Fig. 3a), with the radioactive end positioned at the lower part within the cylinder shaft. Source 2, also a ^{60}Co source with an activity of 185 kBq, was enclosed in a metal foil and secured at the bottom of a metal cup using a circlip. A wire mesh provided protection for the foil located at the front of the cup, and the source was equipped with a 4 mm diameter handle for easy handling (Fig. 3b). Additionally, the sources commonly featured stamped or engraved information about the radionuclide and its initial activity (Fig. 4). This physical characteristic can serve as a valuable signature in nuclear forensic investigations. Figure 4a clearly shows the presence of corrosion on the surface of the ^{241}Am cup-style source. Additionally, the surface of the cup-style ^{60}Co source exhibited green carbonate rust, a clear indicator of ongoing corrosion. Corrosion could provide crucial information about the source condition and historical context, making it a relevant aspect of forensic investigations. Consequently, other methods, such as XRF techniques, might be needed to characterize the corrosion product.

Table 1. Physical Characteristics of sealed radioactive sources.

Characteristic/ Forensic Signature	Source 1	Source 2	Source 3	Source 4
Isotope Name	^{60}Co	^{60}Co	^{137}Cs	^{241}Am
Nominal Activity	74 kBq	185 kBq	185 kBq	185 kBq
Geometry	Cylindrical design	Cup-style design	Disc	Cup-style design
Serial Number	NR-HA 992	None	99 AM 3202783	99AM 320
Labeling Position	Upper and Bottom	Mark on Wooden Box	Marked on the disk	Mark on Wooden Box
Capsule Material	Stainless steel capsule	Mounted within a metal holder and has wire mesh across the open end	Plastic disk	Stainless steel capsule
Container Type	Aluminium Shielded Container well	Small lead castle with lead lid in Wooden Box	Plastic Container	Small lead castle with lead lid in Wooden Box
Outside Capsule Dimensions, diameter x length or thickness	11 mm x 85 mm	14 mm x 22 mm	4.5 mm x 25 mm	4.1 mm x 10 mm
Manufacturer	PHYWE	Griffin & George Ltd	The Nucleus	Griffin & George Ltd
Reference Date of Activity	None	None	None	None
Colour of the Source	Silver	Silver	Orange	Gray
Weight	25.8 g	7.07 g	2.7 g	1.9 g

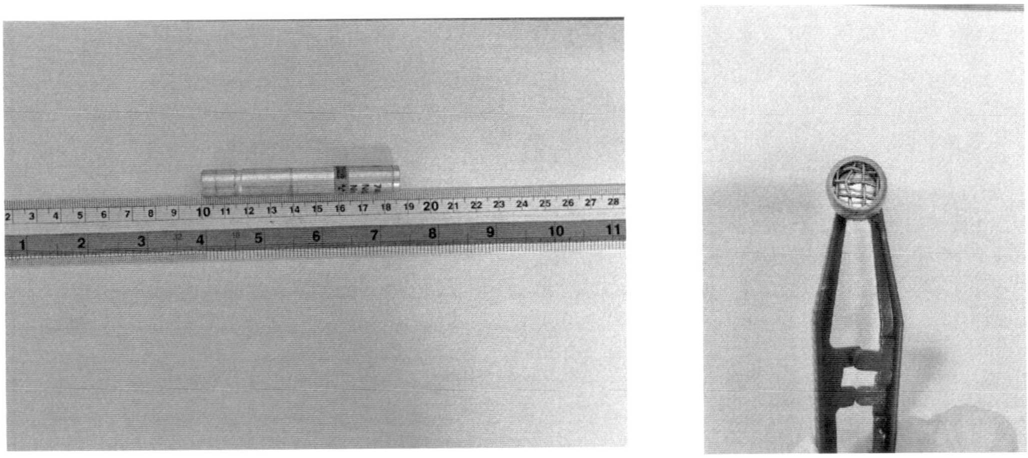

(a) (b)

Figure 3. (a) A cylindrical ^{60}Co source of the 74 kBq type pulled out of its metal holder and (b) a 185 kBq ^{60}Co source enclosed in a metal foil disc.

(a) (b)

Figure 4. Physical attributes, such as embossing and engraving on the source, and corrosion on the surface of the source, as signatures in nuclear forensic investigations.

3.2. Gamma spectrum evaluation

Gamma-ray spectra analysis revealed clear peaks for each gamma source, indicating good resolution (Fig. 5 to 8). The peaks for the cylindrical ^{60}Co and ^{137}Cs disk sources (200-500 keV) were well-defined, allowing for easy identification and differentiation of energy levels. In contrast, the cup-style ^{241}Am source exhibited lower resolution with broader and less defined peaks. This information is crucial for accurate identification and analysis in nuclear forensic investigations. Notably, within the energy range of 200-500 keV, a scattering peak was observed in the cylindrical ^{60}Co source (Fig. 5a and 6a) and the ^{137}Cs disk source (Fig. 7a), but not in the cup-style ^{241}Am source (Fig. 8a). These observations indicate that the source design can influence the observed spectrum. Moreover, the presence of the Compton edge was detected in both the ^{137}Cs plastic disc source (1321.3 keV) (Fig. 7a) and the ^{241}Am cup-style source (125.3 keV) (Fig. 8a), although the energies at which they occurred showed slight variations between the two sources.

The gamma spectra of the ^{60}Co, ^{137}Cs, and ^{241}Am sources unveiled additional peaks not originating from these sources. These unexpected peaks can be attributed to various factors in nuclear spectroscopy. The unusual non-Gaussian shapes of these peaks often point to the influence of impurities or background radiation. In Source 1 with ^{60}Co, significant counts per second (cps) at 1115.1 keV (0.090) and 310.5 keV (0.146) suggest the presence of additional gamma-ray emissions. Similarly, Source 3 with ^{137}Cs exhibits notable cps at 471 keV (0.248), 473.5 keV (1.091), and 477 keV (0.660), indicating unexpected peaks. Additionally, Source 4 with ^{241}Am shows significant cps at 101.1 keV (0.110). The potential presence of impurities within the sources is a common phenomenon in radioactive materials produced through irradiation in a nuclear reactor [3]. These impurities can originate from various sources, including raw materials, accidental contamination during the manufacturing process, or intentional additives aimed at modifying material properties or providing support. The trace elements introduced through these means, along with their activated products and decay progeny, contribute to the observed gamma-ray peaks, creating a distinct signature that reflects a combination of the target material and the reactor [3]. Further analysis should be conducted to confirm the presence of impurities and radionuclides in these sources, as demonstrated by Apostol et al. who observed the existence of ^{241}Am and ^{154}Eu in californium neutron sources using high-resolution gamma spectrometry [9].

IOP Conf. Series: Materials Science and Engineering 1308 (2024) 012016 doi:10.1088/1757-899X/1308/1/012016

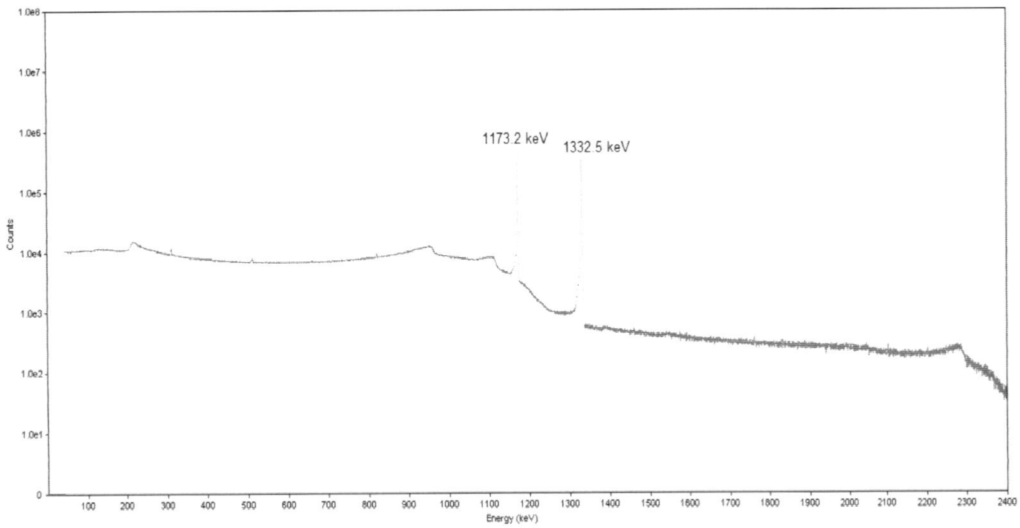

(a)

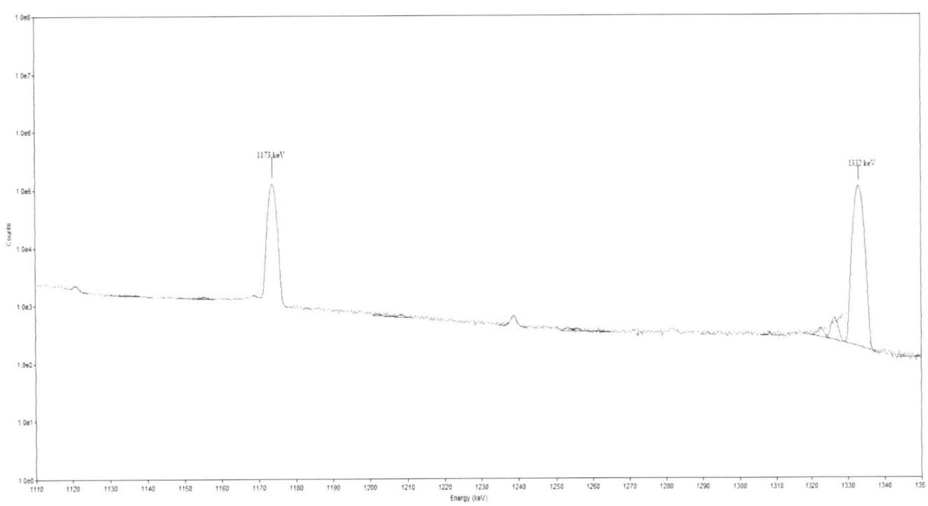

(b)

Figure 5. Gamma-ray spectrum of the ^{60}Co source (74 kBq) at 1173 keV and 1332 keV and b) close-up spectrum

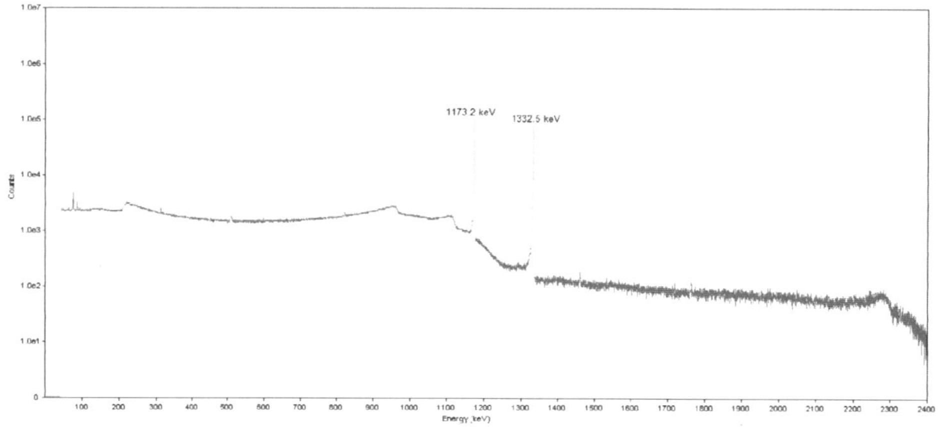

(a)

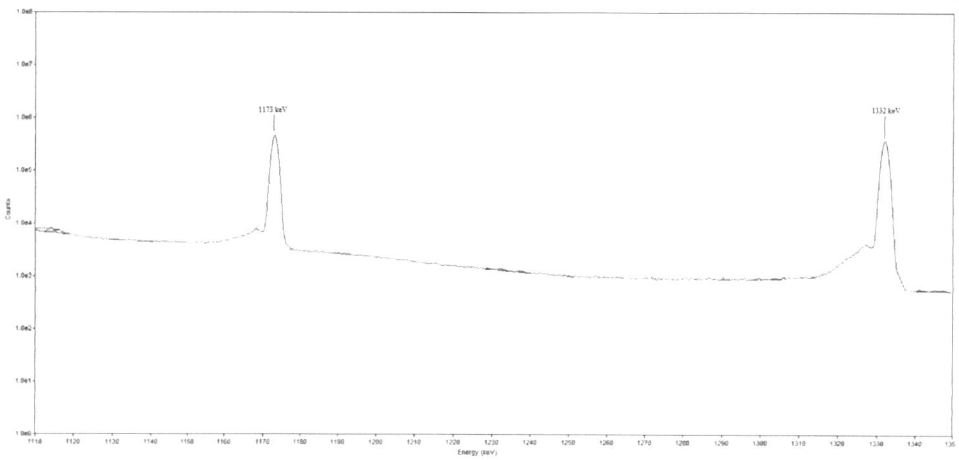

(b)

Figure 6. Gamma-ray spectrum of a) the ^{60}Co source (185 kBq) at 1173 keV and 1332 keV and b) close-up spectrum

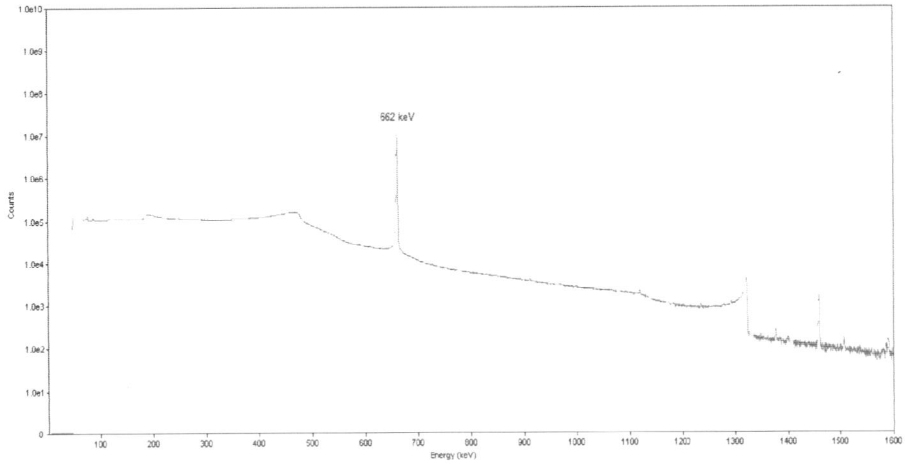

(a)

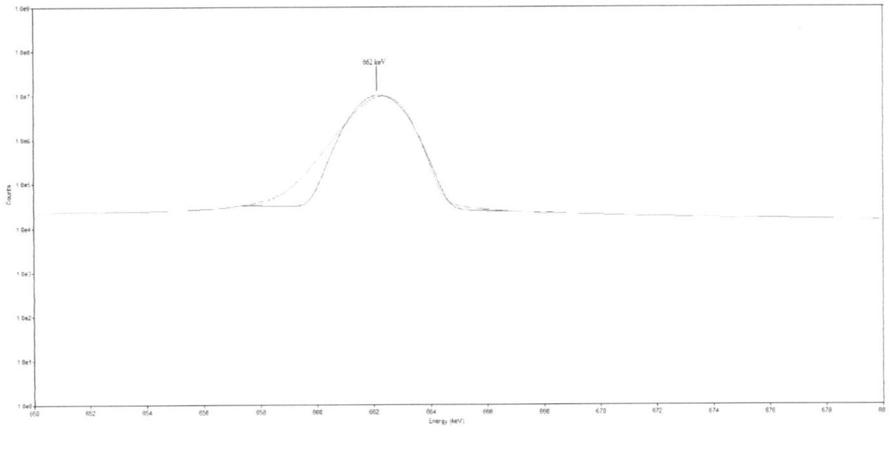

(b)

Figure 7. Gamma-ray spectrum of the ^{137}Cs source (185 kBq) at 662 keV and b) close-up spectrum.

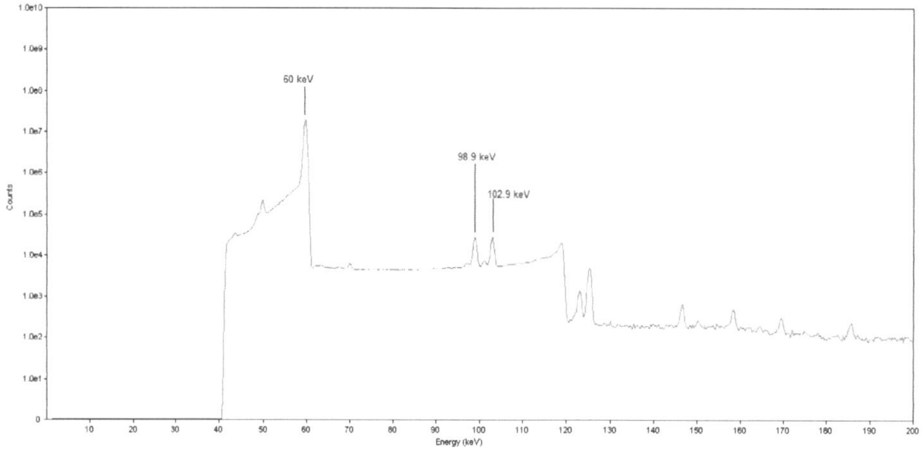

(a)

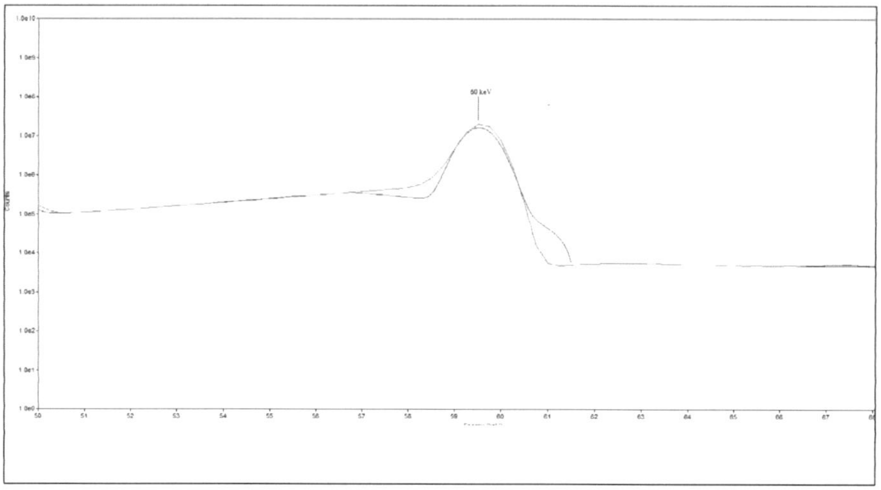

(b)

Figure 8. (a) Gamma-ray spectrum of the ^{241}Am source (185 kBq) at 60 keV and (b) close-up spectrum

3.3. *X-Ray imaging*

X-ray radiography provides the information about the physical characteristics of nuclear materials, including their density, uniformity, and internal structure. Fig. 9 (a) provides valuable information about the internal structure of the disc-type radioactive sources. The X-ray image accurately portrays the location and active area of the ^{241}Am and ^{60}Co radioactive sources within the discs. However, the image does not successfully display these attributes for the ^{137}Cs source. The adjustments made in terms of energy source and exposure time were likely influenced by factors such as the absorption properties of the different radioactive materials and the sensitivity of the X-ray imaging system employed. These variables can impact the optimal settings required to capture clear images of the radioactive sources. Fig.9 (b) presents the image of the ^{60}Co cylindrical source in the aluminium shield, captured at 80 kV with an exposure time of 0.5 min. The internal structure of the source was clearly identifiable, and it was positioned within a rectangular space.

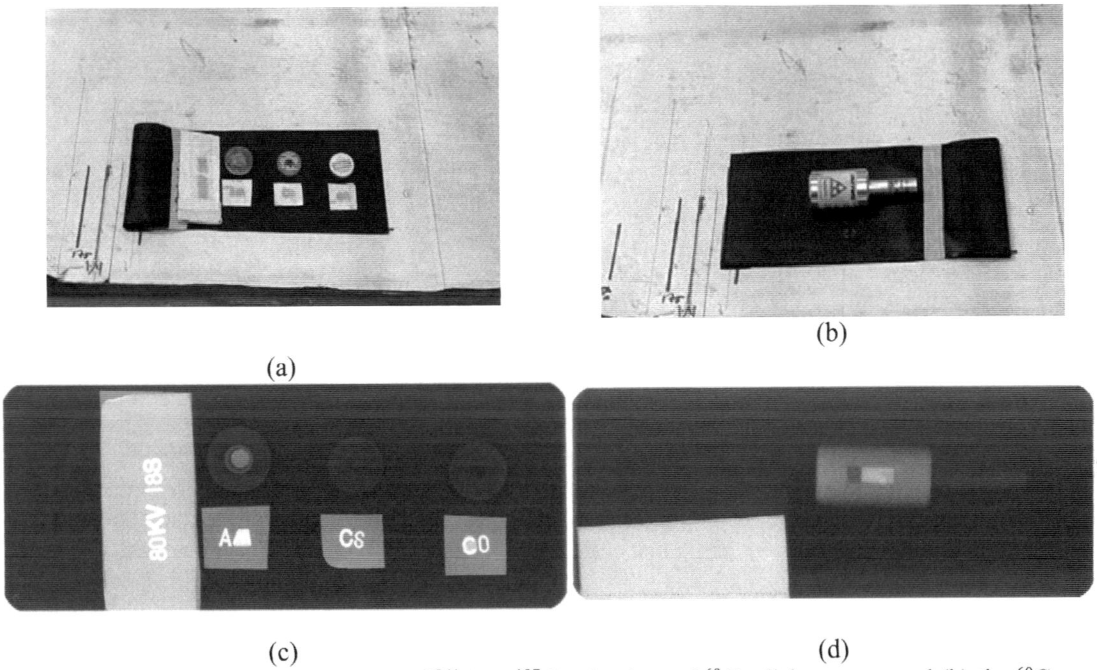

Figure 9. (a) X-ray imaging setup of ^{241}Am, ^{137}Cs plastic, and ^{60}Co disk sources and (b) the ^{60}Co cylindrical source within an aluminum shield, along with (c) and (d) representing the X-ray images of the radioactive sources.

4. Conclusion

In summary, the physical and radiological attributes of sealed radioactive sources, specifically ^{60}Co, ^{137}Cs, and ^{241}Am, provide essential characteristics for their identification, with signatures such as construction, shielding types, and engraved details playing a key role in differentiation. Gamma-ray spectral analysis reveals unique characteristics such as scattering peaks and non-originating peaks, indicating the influence of source design and potential impurities, as reflected in elevated counts per second (cps) values. X-ray radiography enhances the investigation by providing valuable insights into the uniformity and internal structure of sealed radioactive sources. Together, this comprehensive analysis contributes significantly to the establishment of a National Nuclear Forensic Library (NNFL) and enhances Malaysia's capability to identify and analyze radioactive sources that are out of regulatory control. Future research is required to assess the viability of alternative methods for characterizing sealed radioactive sources for nuclear forensic purposes.

5. References

[1] Keegan E, Kristo M J, Toole K, Kips R, Young E 2015 Nuclear Forensics: Scientific analaysis supporting law enforcement and nuclear security investigations. *Analyt. Chem.* **88**(3)1496-1505.

[2] IAEA 2015 Nuclear forensic in support investigation *IAEA Nuclear Security Series No. 2-G (Rev. 1)* (Vienna: International Atomic Energy Agency).

[3] Goldberg M, Finck M, Chamberline D and Carney K 2011 Forensic signature of radiological

IOP Conf. Series: Materials Science and Engineering 1308 (2024) 012016 doi:10.1088/1757-899X/1308/1/012016

sources *52nd Ann. Meeting of the Inst.of Nucl. Mater. Manag. 2011 (INMM52)* (Palm Desert, California, USA).

[4] Chamberlain D, Canaday J, Tang Y, Morman J, Steab J, Naik S, Tsai Y, Sullivan V, Carney K, Finck M Cummings D, Gigillio J and Summers J 2014 Resources and forensics signatures to help determine the origin of sealed radiological sources *Int. Forensics Conf. on Adv. in Nucl. Forensic 7-10, July 2014* (Vienna, Austria) AEA-CN-218-64.

[5] Gregor, J. Kesten, E.A., Kroeger, M., Baron, M. and Eisheh, J.T 2018 Nuclear forensics information obtained from the γ-ray spectra of ^{252}Cf sources of differing provenance. *Nuclear Inst. And Methods in Physic Reasearch B.* **437** 13-19.

[6] Cummings D G, Sommers J D, Adamic M L *et al.* 2009 Characterization of sealed radioactive sources: uncertainty analysis to improve detection methods. *J. Radioanal. Nucl. Chem.* **282** 923–928.

[7] IAEA 2018 Development of a national nuclear forensic library: a system for the identification of nuclear or other radioactive material out of regulatory control (IAEA, Vienna, Austria).

[8] Elias M S, Ibrahim K S, Rahman S A and Wo, MY 2018 Assesment of toxic elements in sediments of Linggi Rver using NAA and ICP-MS techniques *MethodX* **5** 454-465.

[9] Apostol A I, Zsigrai, J, Bagi J, Brandis M, Nikolov J and Mayer K 2019 Characterization of californium sources by gamma spectrometry: relevance for nuclear forensics. *J Radioanal. Nucl. Chem.* **321** 405–412.

Acknowledgments

The authors would like to thank International Atomic Energy Agency (IAEA) for the financial support from CRP-IAEA Research Contract No: 23096 and also would like to express appreciations to the management of Malaysian Nuclear Agency for their continuous assistance, advices and supports on this project.

IOP Conf. Series: Materials Science and Engineering 1308 (2024) 012017 doi:10.1088/1757-899X/1308/1/012017

Evaluation of elemental pollution in soil samples from Klang industrial area

Md Suhaimi Elias[1,a)], Muhammad Azfar Azman[1], Jeremy Andy Dominic Daung[2], Azian Hashim[2], Siti Aminah[1], Nazaratul Ashifa Abdullah Salim[1], Shakirah Shukor[1] and Zalina Laili[1]

[1]Analytical Chemistry Application Group (ACA), Waste and Environmental Technology Division, Malaysian Nuclear Agency, Bangi, 43000 Kajang, Selangor, Malaysia.
[2]Environmental Tracer Application Group (E-TAG), Waste and Environmental Technology Division, Malaysian Nuclear Agency, Bangi, 43000 Kajang, Selangor, Malaysia.

[a]mdsuhaimi@nm.gov.my

Abstract. Rapid development in industries will contribute to the pollution of the soil. Studies of elemental pollution in the soil in the Klang industrial area were conducted to determine the concentration level in that area. Analysis of the content of elemental pollution in soil samples has used neutron activation analysis (NAA) techniques. The average concentration of major elements (Fe, Mg, Ti), heavy metals (As, Sb, Cr, Zn), uranium (U) and thorium (Th) in the soil of the Klang industrial area exceeded the average concentration in granitic igneous rocks. In comparison, the average concentration of rare earth elements (REEs) in soil samples in the study area is lower than the reference value of granitic igneous rocks. The elements of As, Zn and Cr showed a high concentration in the SL 11 area, where business activities, vehicle workshops and residential were located. The REEs show a high concentration in the SL 14 area compared to other sites. The geo-accumulation (Igeo) index was applied to identify the pollution status. The Igeo index for the elements of Mg and U can be categorised as unpolluted to moderately polluted. As and Sb can be categorised as unpolluted to moderately polluted, and as extremely polluted in the areas of SL 11 and SL 14. The element of Th can be categorised as unpolluted to moderately and strongly polluted. The Igeo index of REEs can be categorised as unpolluted in the soil of the Klang Industrial area.

1. Introduction

In Malaysia, rapid development and urbanisation have become issues related to the release of major elements and heavy metals into the environment. Actinide elements (U, Th) become an issue due to their radiation level, and rare earth elements (REEs) activities such as mining, processing, recycling, waste treatment, and disposal are also an issue [1]. In general, sources of major elements, heavy metals, actinides and REEs contamination in soil mainly originate from two sources, namely anthropogenic activities and geogenic processes. The major activities that contribute to anthropogenic pollution such as industrial waste, mining, logging, land clearing, road construction, processing and manufacturing, domestic waste, sewage, agriculture (crops, palm oil and rubber), fuel burning, livestock farming and shipping activities [2–4]. Meanwhile, the geogenic process occurs with weathering of soil and rocks, soil erosion, terrestrial runoff and atmosphere deposition[3,4]. Studies of heavy metal contamination in the soil of Malaysia have been done by various agencies and locations [5–9]. For instance, studies of the major elements in soil samples collected in Gebeng, Pahang, indicate that the Fe element is higher compared to other elements. Meanwhile, the concentrations of heavy metals (As and Sb) in soil samples of the Gebeng industrial area were slightly higher than the continental crust average [5]. The additional elemental pollution (Fe, As, Sb) in the soil of the Gebeng industrial area is due to the receiving of anthropogenic pollution from activities such as land clearing, road and drainage constructions and the opening of new industrial areas [5].

Studies of REEs showed that the concentration of REEs in the soil of the Gebeng industrial area is slightly lower compared to the continental crust average concentrations [5]. The concentration of heavy metals (Cu, Cr, Ni, Cd, Pb) in soil collected around Perlis indicates the heavy metals' contamination distribution due to industrialisation, urbanisation and agricultural activities [6,7]. A study of elements (As, Dy, Lu, Ta, Tb, U, Zn) in soil collected from Manjung and central Perak shows significant enrichment elemental concentration of As, Dy, Lu, Ta, Tb, U, and Zn. The possible sources of element pollution originated from the usage of pesticides and fertilisers. The pesticide and overuse of fertiliser in agriculture are widely contributing to soil pollution in the areas of Manjung and central Perak [8]. Other studies related to the RREs pollution in soil samples collected from Lahat, Perak, indicate that the sources of REEs pollution originated from industrial and mining areas[9].

The objective of this study is to determine the concentration, distribution and contamination level of major elements, heavy metals, actinides and REEs in the soil sample of the Klang Industrial area. The study of major elements (Fe. Mg, Ti), heavy metals (As, Sb, Cr, Zn), actinide (U, Th), and REEs (La, Ce, Nd, Sm, Eu, Tb, Dy, Yb, Lu) contamination levels in the soils is vital due to their capabilities to be uptake by plants, vegetation and crops. Major elements, heavy metals, actinides and rare earth elements can be transferred through the food chain and could have an adverse effect on human health at high concentrations [7,8]. In addition, the significance of this study will provide crucial information to identify specific elements and the sources contributing to major pollution in this area.

2. Methodology

2.1. Sampling of soil samples from the Klang industrial area

Fourteen locations in the Klang Industrial area were selected for the sampling of soil samples. The soil samples were collected at 0 – 15 cm depth using an auger and transferred into a polyethylene bag [1]. Soil samples were then transported to the laboratory and dried at 60°C in an oven, followed by grounding using agate mortar[5]. The powdered soil samples were kept in polyethylene bottles for analysis. The soil sampling location and the description of the activities at the sampling site of Klang industrial areas are depicted in Figure 1 and Table 1, respectively.

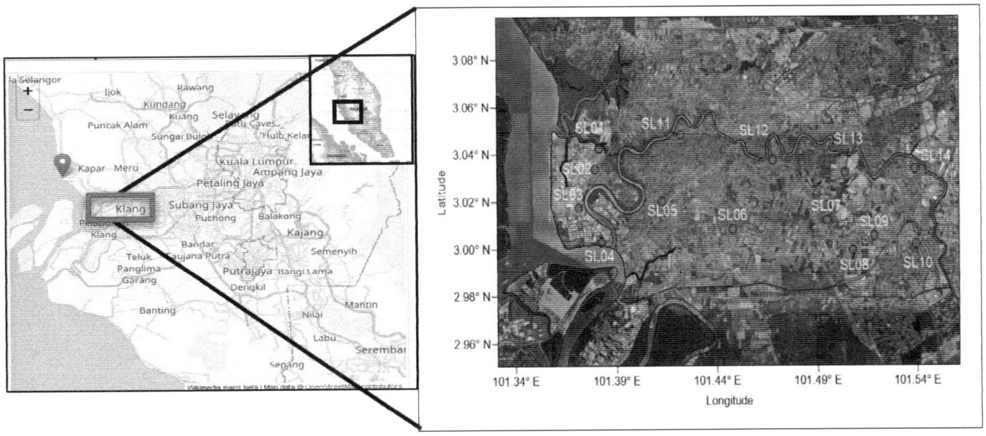

Figure 1. Sampling locations of soil samples collected from the Klang Industrial area.

Table 1. The description of activities in the sampling site.

Location	Activities description
SL01 SL02 SL03	Sultan Suleiman Industrial Area – Clothes and fabric, electronics, pharmaceutical, manufacturing, chemical, rubber product, food industries, fertiliser industries
SL04	Klang North Industrial Area – Steel industries, electronic, fabric, rubber products, manufacturing, motor and machine industries.
SL05	Raja Uda Area- Residential and shop lot
SL06	Taman Klang Jaya – Residential area, shop lot, chemical, manufacturing filter, food and beverage industries
SL07	AMJ Industrial Park – Plastic, manufacturing, food, steel, motor and machine industries
SL08	Techno Industrial Park – Workshop, concrete, manufacturing, plastic, motor and machine industries.
SL09	Seksyen 32 – Residential area, rubber product, recycling paper and plastic industries.
SL10	Kota Kemuning Industrial Area – Chemical, electronic, paint, rubber industries.
SL11	Residential, shop lots and vehicle workshop
SL12	Residential
SL13	Residential (Control area)
SL14	Axis Industrial Park – Food and beverage manufacturers, rubber, recycling centre, paper, chemical, plastic industries.

2.2 Analysis of soil samples by using the neutron activation analysis (NAA) technique

Approximately 0.20 grams of homogenous powdered soil samples were weighed and placed into polyethylene vials. The soil samples, standard reference materials (SRM) (San Joaquin Soil – NIST SJS – 2709a and Sediment Lake – SL-1) and blank were irradiated for six hours at the PUSPATI TRIGA Research Reactor facility of the Malaysian Nuclear Agency and allowed for a decay process of two to four days, followed by one hour of counting. The counting process of the irradiated soil samples, SRM, and blank was performed using a gamma spectrometer. The net peak areas (gamma ray intensity) of the major elements, heavy metal, actinides and REEs in the soil samples were compared to the SRM (NIST SJS – 2709a) as the standard comparator for concentration measurement. Meanwhile, the Sediment Lake – SL-1 was used as quality control and quality assurance (QA/QC). The Calculations of major elements, heavy metal, actinides and REEs concentrations in soil samples were performed according to Equation (1)[10–12]

$$Csmp = Cstd \times (Wstd \times Nsmp)/(Wsmp \times Nstd) \quad (1)$$

Where: Nsmp and Nstd are the net counts of the peak area of the soil sample and standard, respectively; Csmp and Cstd are the concentration of the soil sample and standard, respectively; Wsmp and Wstd are the weight of the soil sample and standard, respectively.

2.3 Contour map

The contour maps of major elements, heavy metal, actinides and REEs concentration were generated using Surfer software. Kriging interpolations were applied to illustrate the contour map of major elements, heavy metals, actinides and REEs concentrations in the Klang Industrial area.

3. Results and discussion

3.1 Concentration of major elements, heavy metals, actinide and REEs in the soil of the Klang Industrial area

Fourteen locations in the Klang industrial area were selected for the sampling of the soil samples. The elemental content in the soil samples was analysed using the Neutron activation analysis (NAA) technique. The concentrations of major elements, heavy metals, actinide (U, Th) and REEs in the soil of the Klang industrial area and other locations in Malaysia were tabulated in Table 2. The selection of reference crust (igneous granitic rock) referred to most of the elements from the normalised (ratio) result of average concentration value in the soil of Klang industrial area to the igneous granitic rock values are close to 1. The Fe, Mg, and Ti concentrations were ranged from 3203 – 59205 mg/kg, 1093 – 38906 mg/kg, and 1013 – 9000mg/kg, respectively. The average concentrations of major elements (Fe, Mg and Ti) showed higher concentrations compared to the igneous granitic rock crust of those elements. This indicates additional of Fe, Mg and Ti concentrations in the soil of the Klang industrial area. The elemental concentrations of As, Cr, Zn and Sb were ranged from 3.8 – 428.7, 5.5 – 190.5, 14.1 – 866.6 and 0.5 – 18.5mg/kg, respectively. The average concentrations of heavy metals (As, Cr, Zn and Sb) were also higher than the igneous granitic rock of the respective elements. A similar trend was observed for the actinide elements (U, Th), where the concentrations of U and Th were ranged from 0.59 – 10.21 and 2.7 – 64.9mg/kg, respectively. The average concentration of U and Th in soil samples of the Klang industrial area was higher compared to the igneous granitic rock (IGR) values of those elements (Table 2). This indicates the additional and accumulation of major elements, heavy metals and actinide elements in the soil of the Klang industrial area. The possible sources of major elements, heavy metals and actinide pollution likely originated from anthropogenic activities and natural land-based (landslides and terrestrial runoff). The concentration of rare earth elements (REEs) such as La, Ce, Nd, Sm, Eu, Tb, Dy, Yb, and Lu were ranged from 9.3 – 71.4, 17.3 – 131.3, 6.6 – 36.8, 1.07 – 6.78, 0.16 – 0.94, 0.16 – 1.23, 0.94 – 4.44, 0.29 – 3.89, and 0.11 – 0.76mg/kg, respectively (Table 2). However, average concentrations of all REEs showed lower than igneous granitic rock values of the respective elements. This clearly indicates no additional REEs concentrations in the soil samples of the Klang industrial area.

Table 2(a): Concentration of major elements, heavy metals, actinides (U, Th) and rare earth elements (REEs) (mg/kg d.w) in the soil of Klang Industrial area and other area in Malaysia.

Location	Fe	Mg	Ti	As	Cr	Zn	Sb	U	Th
SL01	3203	1286	1013	3.8	5.5	28.6	0.5	0.59	2.7
SL02	17481	4213	3332	18.4	38.0	147.5	1.4	5.18	13.7
SL03	13457	1093	2376	37.4	24.4	318.0	2.9	3.59	11.4
SL04	34966	3288	3734	105.2	52.1	170.3	17.3	3.60	15.8
SL05	31656	2730	3079	61.5	45.0	557.9	3.9	4.96	14.1
SL06	35265	1754	4700	99.0	54.7	83.7	3.4	2.91	15.4
SL07	59205	7704	4276	109.5	74.9	866.6	3.0	3.51	23.5
SL08	35006	7688	3877	157.5	73.2	852.1	4.1	6.47	22.3
SL09	52196	5327	5852	67.7	25.9	57.8	1.6	7.77	62.2
SL10	33821	21924	6015	21.2	52.2	41.9	2.3	10.21	64.9
SL11	26732	18115	4073	428.7	76.5	121.7	15.2	4.06	24.9
SL12	21845	9573	4651	43.6	130.5	14.1	10.1	2.88	15.8
SL13	36747	14192	3898	71.2	64.7	25.7	4.0	2.54	24.1
SL14	45347	38906	9000	59.1	190.5	57.0	18.5	6.04	43.8
Average	31923	9842	4277	91.7	64.9	238.8	6.3	4.59	25.3
Igneous granitic rock[13]	29600	9400	3400	1.9	22	60	0.2	3.00	8.5
Manjung and central Perak N =26$ [8]	14500	10700	3607	18.0	45.9	119	1.04	11.7	24.0
Gebeng, Pahang, N = 27@ [5]	53097	-	-	7.30	112.8	40.9	0.45	3.09	15.9

$ and @ =Average concentration of elements; N = Number of sampling locations

Table 2(b): Concentration of major elements, heavy metals, actinides (U, Th) and rare earth elements (REEs) (mg/kg d.w) in the soil of Klang Industrial area and other area in Malaysia (continue).

Location	La	Ce	Nd	Sm	Eu	Tb	Dy	Yb	Lu
SL01	9.3	17.3	6.6	1.07	0.16	0.16	0.94	0.29	0.16
SL02	29.9	66.5	31.5	4.60	0.75	0.70	1.94	2.07	0.17
SL03	17.8	38.1	18.7	2.63	0.30	0.50	1.48	1.28	0.22
SL04	22.2	42.7	12.4	3.06	0.56	0.62	3.12	2.46	0.46
SL05	15.8	73.9	20.6	2.91	0.44	0.58	1.63	1.87	0.25
SL06	14.9	29.1	7.1	2.17	0.50	0.50	1.88	2.33	0.48
SL07	23.6	49.0	14.7	2.65	0.52	0.49	1.74	1.40	0.49
SL08	27.1	66.8	36.8	4.29	0.82	0.78	3.94	2.36	0.36
SL09	37.2	64.6	27.3	3.85	0.33	0.62	2.45	0.26	0.11
SL10	41.2	77.2	28.0	5.38	0.49	0.84	2.69	3.89	0.63
SL11	54.4	103.3	35.2	6.78	0.94	1.23	3.05	3.78	0.60
SL12	41.3	51.9	22.3	3.91	0.62	0.89	2.82	3.54	0.55
SL13	25.7	44.6	16.8	2.96	0.50	0.80	1.89	2.90	0.53
SL14	71.4	131.3	19.6	6.58	0.91	1.05	4.44	3.36	0.76
Average	30.8	61.2	21.2	3.77	0.56	0.70	2.43	2.27	0.41
Igneous granitic rock[13]	45	81	33	8.80	1.40	1.40	6.30	3.50	1.10
Manjung and central Perak N =26$ [8]	36.0	80.3	-	6.10	0.64	2.42	6.77	7.39	0.60
Gebeng, Pahang, N = 27@ [5]	20.1	48.9	16.8	3.00	0.45	0.41	-	2.05	0.26

$ and @ =Average concentration of elements; N = Number of sampling locatio

3.2 Geo-accumulation index (I_{geo}) in the soil of the Klang industrial area

The geo-accumulation index (I_{geo}) was used to identify the degree of anthropogenic contamination by comparing the current concentration status of elemental contamination over the pre-industrial levels of contamination in soil or sediment[12,14–16]. The geo-accumulation index (I_{geo}) was initially introduced by Müller in 1969[17]. The geo-accumulation index can be calculated using Equation (2). I_{geo} can be classified into seven classifications to determine the level of soil contamination. The soil contamination classification are as follows: I_{geo} value < 0 = uncontaminated (I_{geo} class 0), I_{geo} value > 0 to 1 = uncontaminated to moderately contaminated (I_{geo} class 1), I_{geo} value > 1 to 2 = moderately contaminated (I_{geo} class 2), I_{geo} value > 2 to 3 = moderately to strongly contaminated (I_{geo} class 3), I_{geo} value > 3 to 4 = strongly contaminated (I_{geo} class 4), I_{geo} value 4 to 5 = strongly to extremely contaminated (I_{geo} class 5) and I_{geo} value > 5 = extremely contaminated (I_{geo} class 6)[14–16,18,19]. Major elements, heavy metals, actinides and REEs contamination in the soil of the Klang industrial area were compared to the igneous granitic rock-IGR crust values for the calculation of the geo-accumulation index[13,20].

$$I_{geo} = log_2\left(\frac{C_n}{1.5\times B_n}\right) \tag{2}$$

where: C_n is the measured concentration of the interest element, B_n is the background value of interest reference element concentration from crust value (igneous granitic rock - IGR crust) [13] and 1.5 is a correction factor for the variation of the background values due to lithogenic effects in the soil.

I_{geo} values and I_{geo} class of the major elements, heavy metals, actinide and REEs in the soil samples of the Klang industrial area were tabulated in Table 3. The status of the geo-accumulation index (I_{geo}), the jitter and whisker plot of the I_{geo} value of major elements, heavy metals, actinides and REEs contamination in the soil of Klang industrial area are depicted in Figure 2. The lower border of the horizontal line illustrates the 25 percentile value, and the 75 percentile value is the top border of the horizontal line of the box, whilst the horizontal line shows the median value in the box plot.

The I_{geo} values of Fe, Mg and Ti were ranged from less than 0 to 1.5 (I_{geo} class 0 to 2). Generally, the major elements in the soil of the Klang industrial area can be categorised as uncontaminated to moderately contaminated. The elements of Fe and Ti can be categorised uncontaminated to as uncontaminated to moderately contaminated, while Mg can be categorised as uncontaminated to moderately contaminated. I_{geo} values of As and Sb elements were ranged from 0.4 – 7.2 and 0.6 – 5.9, respectively (I_{geo} class = class 1 to class 6). This indicates that the elements of As and Sb in the soil sample of the Klang industrial area can be categorised as uncontaminated to moderately contaminated and as extremely contaminated. I_{geo} values of Cr ranged from – 2.6 to 2.5 (I_{geo} class = class 0 to class 3), indicating that the Cr in the soil sample can be categorised as uncontaminated and as moderately to strongly contaminated. Element of Zn can be classified as uncontaminated to strongly contaminated with I_{geo} values of – 2.7 to 3.3 (I_{geo} class = class 0 to class 4). The actinide element (U, Th) showed slightly higher concentration in the SL10 location. The I_{geo} values of U and Th were ranged from – 2.9 to 1.2 and – 2.2 to 2.3, respectively. The U contamination can be categorised as uncontaminated to moderately contaminated (I_{geo} class = class 0 to class 2), while the Th can be classified as uncontaminated and as moderately to strongly contaminated (I_{geo} class = class 0 to class 3). The I_{geo} values of REEs shown below 0 indicate uncontaminated (I_{geo} class = class 0) except for the element of La and Ce in the location of SL14, with I_{geo} value of 0.10 (I_{geo} class = class 1). The industries should take action to prevent the release of elemental contamination into the soil by regularly monitoring the release of dust from the chimney and discharge from drainage or retention ponds, especially in locations categorised as moderately to extremely contaminated (I_{geo} class = class 2 to class 6). The government agencies and authorities should also regularly monitor the industries to follow the Environment Quality Act 1974.

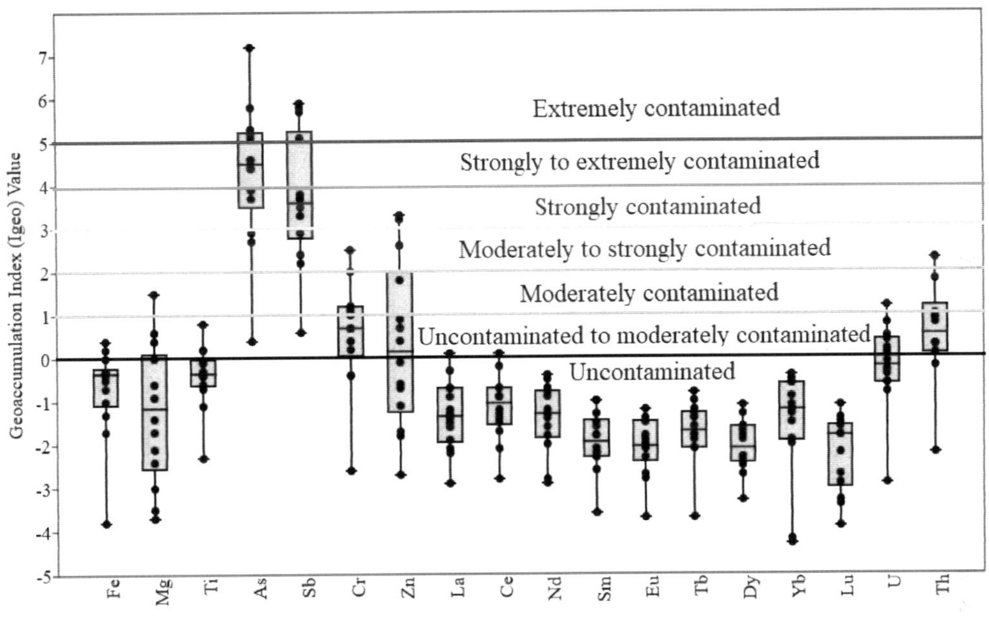

Figure 2. Geo-accumulation Index (I_{geo}) status of major elements, heavy metals, actinide (U, Th) and REEs in the soil of the Klang industrial area.

Table 3(a). I_{geo} values and (I_{geo} Class) of the major elements, heavy metals, actinide (U, Th) and rare earth elements (REEs) in the soil sample of the Klang Industrial area

Location	Fe	Mg	Ti	As	Cr	Zn	Sb	U	Th
				I_{geo} value (I_{geo} class)					
SL01	-3.8 (0)	-3.5 (0)	-2.3 (0)	0.4 (1)	-2.6 (0)	-1.7 (0)	0.6 (1)	-2.9 (0)	-2.2 (0)
SL02	-1.3 (0)	-1.7 (0)	-0.6 (0)	2.7 (3)	0.2 (1)	0.7 (1)	2.2 (3)	0.2 (1)	0.1 (1)
SL03	-1.7 (0)	-3.7 (0)	-1.1 (0)	3.7 (4)	-0.4 (0)	1.8 (2)	3.3 (4)	-0.3 (0)	-0.2 (0)
SL04	-0.3 (0)	-2.1 (0)	-0.4 (0)	5.2 (6)	0.7 (1)	0.9 (1)	5.8 (6)	-0.3 (0)	0.3 (1)
SL05	-0.5 (0)	-2.4 (0)	-0.7 (0)	4.4 (5)	0.4 (1)	2.6 (3)	3.7 (4)	0.1 (1)	0.1 (1)
SL06	-0.3 (0)	-3.0 (0)	-0.1 (0)	5.1 (6)	0.7 (1)	-0.1 (0)	3.5 (4)	-0.6 (0)	0.3 (1)
SL07	0.4 (1)	-0.9 (0)	-0.3 (0)	5.3 (6)	1.2 (2)	3.3 (4)	3.3 (4)	-0.4 (0)	0.9 (1)
SL08	-0.3 (0)	-0.9 (0)	-0.4 (0)	5.8 (6)	1.1 (2)	3.2 (4)	3.8 (4)	0.5 (1)	0.8 (1)
SL09	0.2 (1)	-1.4 (0)	0.2 (1)	4.6 (5)	-0.4 (0)	-0.6 (0)	2.4 (3)	0.8 (1)	2.3 (3)
SL10	-0.4 (0)	0.6 (1)	0.2 (1)	2.9 (3)	0.7 (1)	-1.1 (0)	2.9 (3)	1.2 (2)	2.3 (3)
SL11	-0.7 (0)	0.4 (1)	-0.3 (0)	7.2 (6)	1.2 (2)	0.4 (1)	5.7 (6)	-0.1 (0)	1.0 (1)
SL12	-1.0 (0)	-0.6 (0)	-0.1 (0)	3.9 (4)	2.0 (2)	-2.7 (0)	5.1 (6)	-0.6 (0)	0.3 (1)
SL13	-0.3 (0)	0.0 (0)	-0.4 (0)	4.6 (5)	1.0 (1)	-1.8 (0)	3.7 (4)	-0.8 (0)	0.9 (1)
SL14	0.0 (0)	1.5 (2)	0.8 (1)	4.4 (5)	2.5 (3)	-0.7 (0)	5.9 (6)	0.4 (1)	1.8 (2)

Table 3(a). I_{geo} values and (I_{geo} Class) of the major elements, heavy metals, actinide (U, Th) and rare earth elements (REEs) in the soil sample of the Klang Industrial area (continue)

Location	La	Ce	Nd	Sm	Eu	Tb	Dy	Yb	Lu
				I_{geo} value (I_{geo} class)					
SL01	-2.9 (0)	-2.8 (0)	-2.9 (0)	-3.6 (0)	-3.7 (0)	-3.7 (0)	-3.3 (0)	-4.2 (0)	-3.4 (0)
SL02	-1.2 (0)	-0.9 (0)	-0.7 (0)	-1.5 (0)	-1.5 (0)	-1.6 (0)	-2.3 (0)	-1.3 (0)	-3.3 (0)
SL03	-1.9 (0)	-1.7 (0)	-1.4 (0)	-2.3 (0)	-2.8 (0)	-2.1 (0)	-2.7 (0)	-2.0 (0)	-2.9 (0)
SL04	-1.6 (0)	-1.5 (0)	-2.0 (0)	-2.1 (0)	-1.9 (0)	-1.8 (0)	-1.6 (0)	-1.1 (0)	-1.8 (0)
SL05	-2.1 (0)	-0.7 (0)	-1.3 (0)	-2.2 (0)	-2.3 (0)	-1.9 (0)	-2.5 (0)	-1.5 (0)	-2.7 (0)
SL06	-2.2 (0)	-2.1 (0)	-2.8 (0)	-2.6 (0)	-2.1 (0)	-2.1 (0)	-2.3 (0)	-1.2 (0)	-1.8 (0)
SL07	-1.5 (0)	-1.3 (0)	-1.8 (0)	-2.3 (0)	-2.0 (0)	-2.1 (0)	-2.4 (0)	-1.9 (0)	-1.8 (0)
SL08	-1.3 (0)	-0.9 (0)	-0.4 (0)	-1.6 (0)	-1.4 (0)	-1.4 (0)	-1.3 (0)	-1.2 (0)	-2.2 (0)
SL09	-0.9 (0)	-0.9 (0)	-0.9 (0)	-1.8 (0)	-2.7 (0)	-1.8 (0)	-1.9 (0)	-4.3 (0)	-3.9 (0)
SL10	-0.7 (0)	-0.7 (0)	-0.8 (0)	-1.3 (0)	-2.1 (0)	-1.3 (0)	-1.8 (0)	-0.4 (0)	-1.4 (0)
SL11	-0.3 (0)	-0.2 (0)	-0.5 (0)	-1.0 (0)	-1.2 (0)	-0.8 (0)	-1.6 (0)	-0.5 (0)	-1.5 (0)
SL12	-0.7 (0)	-1.2 (0)	-1.2 (0)	-1.8 (0)	-1.8 (0)	-1.2 (0)	-1.7 (0)	-0.6 (0)	-1.6 (0)
SL13	-1.4 (0)	-1.4 (0)	-1.6 (0)	-2.2 (0)	-2.1 (0)	-1.4 (0)	-2.3 (0)	-0.9 (0)	-1.6 (0)
SL14	0.1 (1)	0.1 (1)	-1.3 (0)	-1.0 (0)	-1.2 (0)	-1.0 (0)	-1.1 (0)	-0.6 (0)	-1.1 (0)

3.3 Contour map for distribution of major elements, heavy metals, actinide (U, Th) and REEs concentration in soil of Klang industrial area

The contour maps of the distribution of major elements, heavy metals, actinide and REEs concentrations in the soil of the Klang industrial area were depicted in Figure 3(a) – (c), Figure 3 (d) – (g), Figure (h) – (i) and Figure 3 (j) – (r), respectively. The major elements of Mg and Ti showed higher concentrations in location of SL 14 compared to other sites. The As element showed higher in SL11, which is located of shop lots, vehicle workshops and residential. The elements of Cr and Sb showed higher in SL14, where the Axis Industrial Park was located, which related to the activities of food and beverage manufacturers, rubber, recycling centre, paper, chemical, and plastic industries. Zn concentration showed slightly elevated in SL07 and SL08 locations. The actinides (U and Th) concentration was higher in the Kota Kemuning industrial area, which was located of the chemical, electronic, paint, and rubber industries (SL10).

The groups of REEs (La, Ce, Sm, Eu, Dy, Tb, Lu and Yb) showed a slightly similar trend on the contour map and higher concentration in SL11 and SL14 locations compared to other locations. However, the concentration of REEs in SL 11 and SL14 locations was considered low compared to the igneous granitic rock (IGR) and the geo-accumulation index's contamination status of REEs in soil samples of the Klang industrial area was classified as uncontaminated.

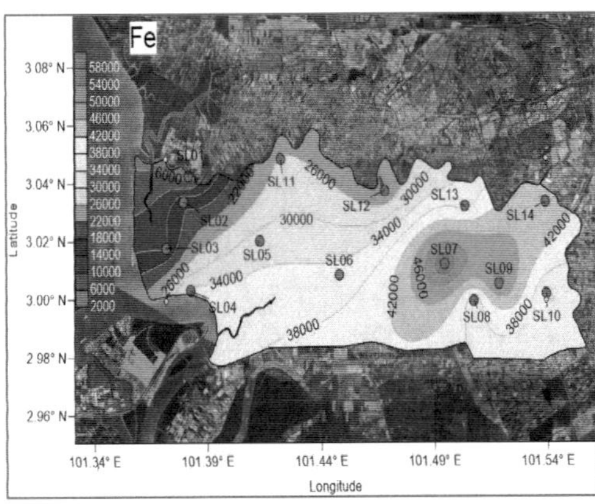

(a)

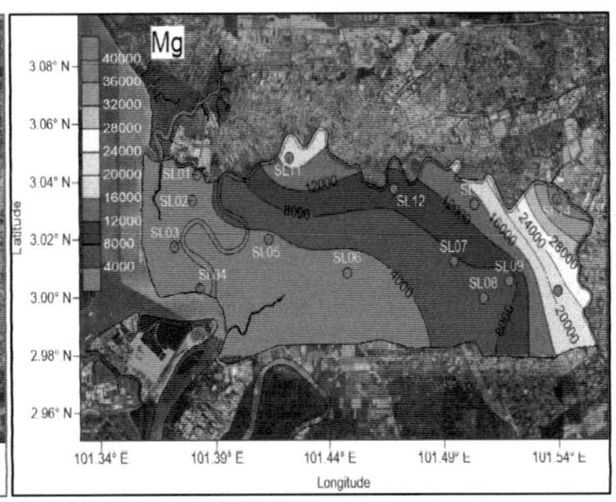

(b)

IOP Conf. Series: Materials Science and Engineering 1308 (2024) 012017 doi:10.1088/1757-899X/1308/1/012017

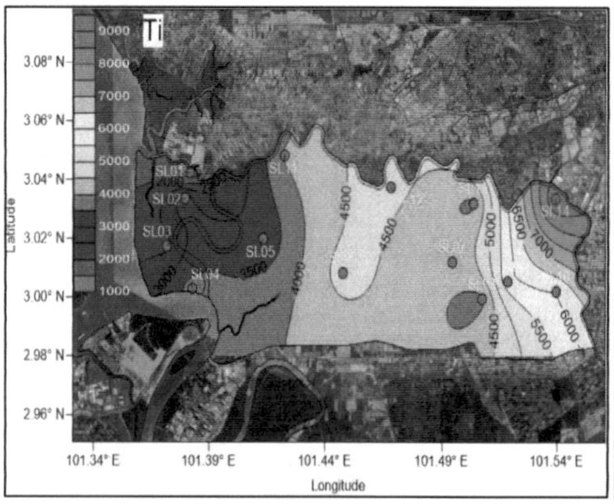

(c)

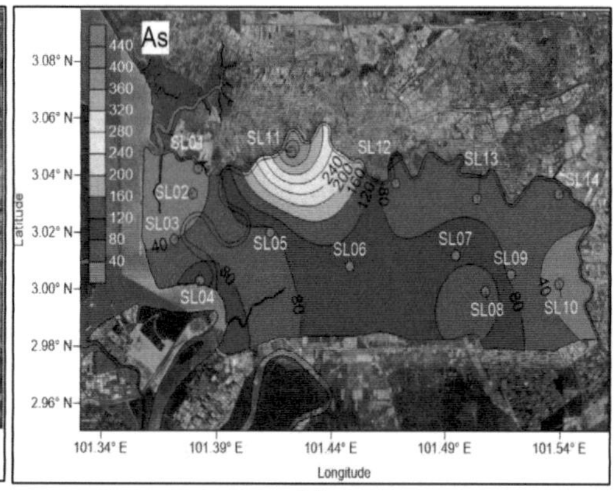

(d)

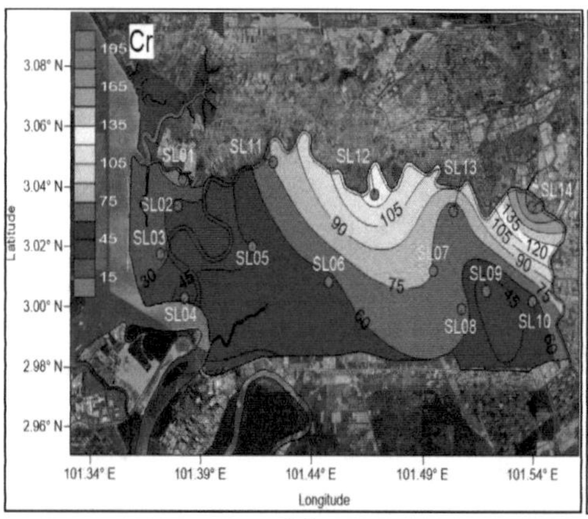

(e)

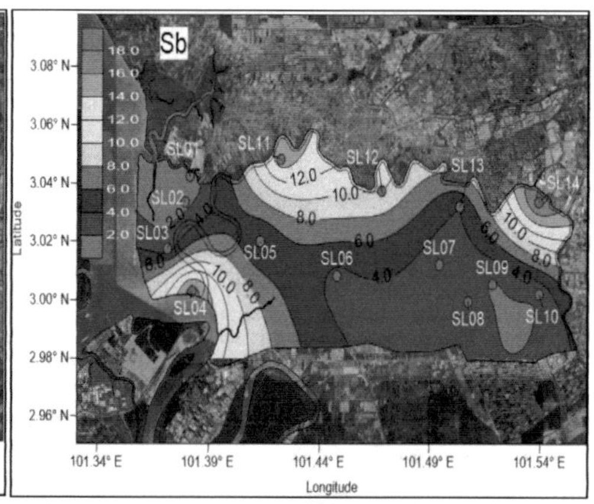

(f)

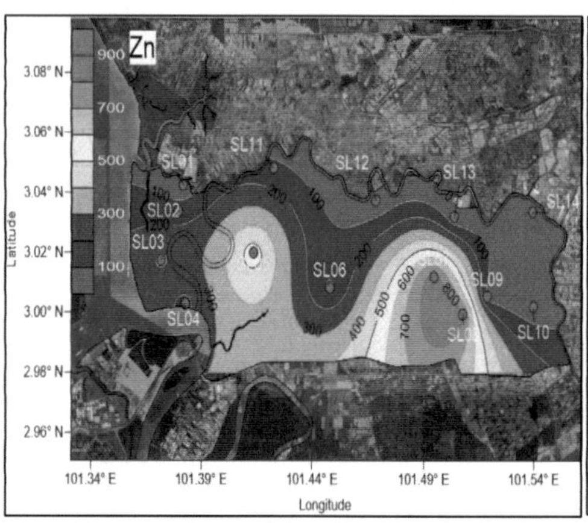

(g)

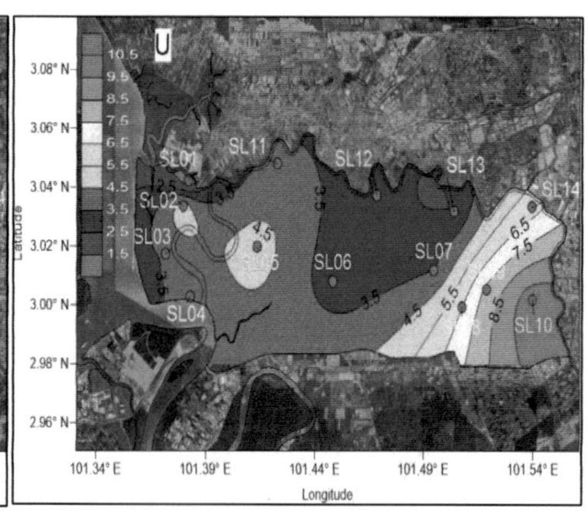

(h)

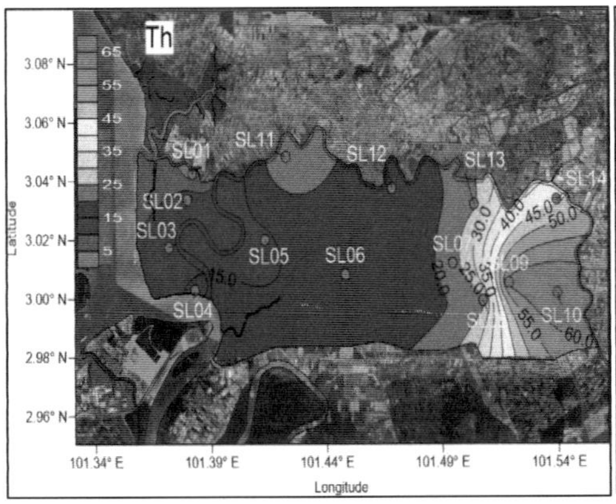
Th
SL01
SL02
SL03
SL04
SL05
SL06
SL11
SL12
SL13
SL14
SL10
3.08° N
3.06° N
3.04° N
3.02° N
3.00° N
2.98° N
2.96° N
101.34° E
101.39° E
101.44° E
101.49° E
101.54° E
Longitude
Latitude

(i)

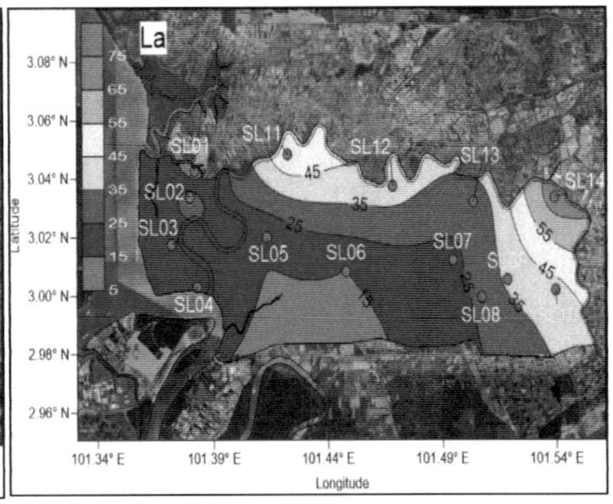
La
SL01
SL02
SL03
SL04
SL05
SL06
SL07
SL08
SL11
SL12
SL13
SL14
3.08° N
3.06° N
3.04° N
3.02° N
3.00° N
2.98° N
2.96° N
101.34° E
101.39° E
101.44° E
101.49° E
101.54° E
Longitude
Latitude

(j)

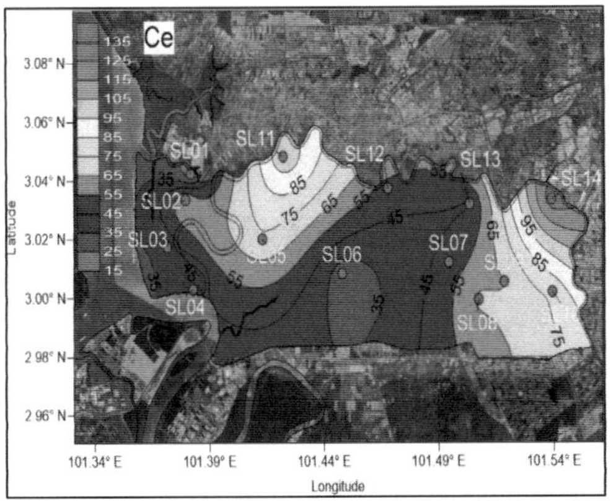
Ce
SL01
SL02
SL03
SL04
SL06
SL07
SL08
SL11
SL12
SL13
SL14
3.08° N
3.06° N
3.04° N
3.02° N
3.00° N
2.98° N
2.96° N
101.34° E
101.39° E
101.44° E
101.49° E
101.54° E
Longitude
Latitude

(k)

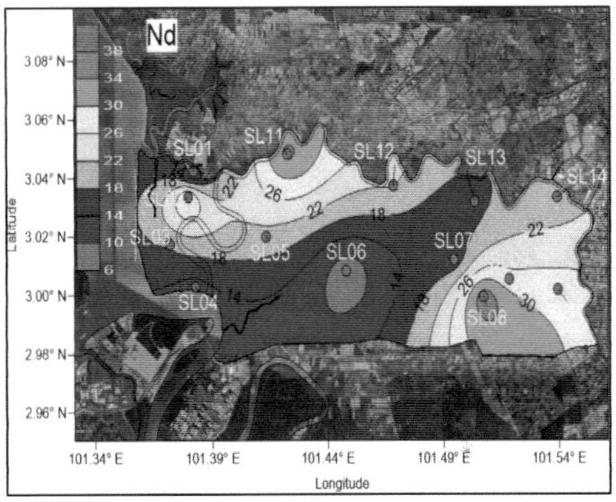
Nd
SL01
SL04
SL05
SL06
SL07
SL08
SL11
SL12
SL13
SL14
3.08° N
3.06° N
3.04° N
3.02° N
3.00° N
2.98° N
2.96° N
101.34° E
101.39° E
101.44° E
101.49° E
101.54° E
Longitude
Latitude

(l)

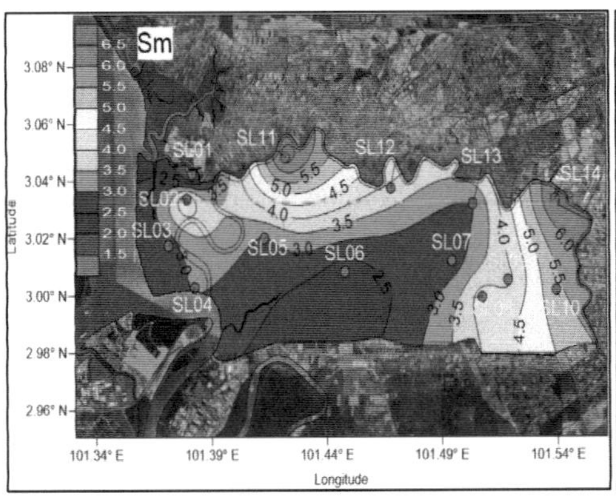
Sm
SL01
SL03
SL04
SL05
SL06
SL07
SL10
SL11
SL12
SL13
SL14
3.08° N
3.06° N
3.04° N
3.02° N
3.00° N
2.98° N
2.96° N
101.34° E
101.39° E
101.44° E
101.49° E
101.54° E
Longitude
Latitude

(m)

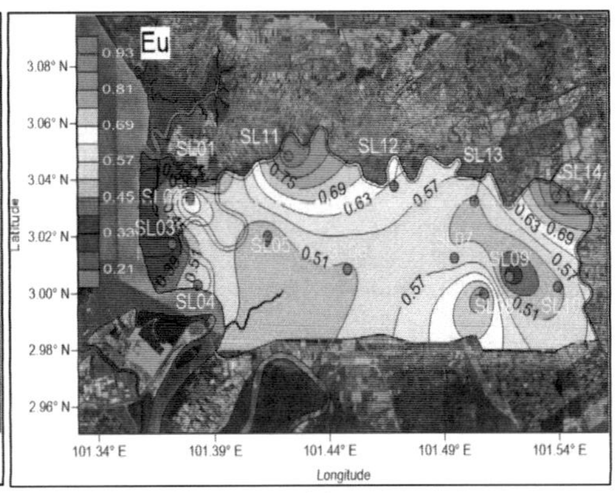
Eu
SL01
SL03
SL04
SL11
SL12
SL13
SL14
3.08° N
3.06° N
3.04° N
3.02° N
3.00° N
2.98° N
2.96° N
101.34° E
101.39° E
101.44° E
101.49° E
101.54° E
Longitude
Latitude

(n)

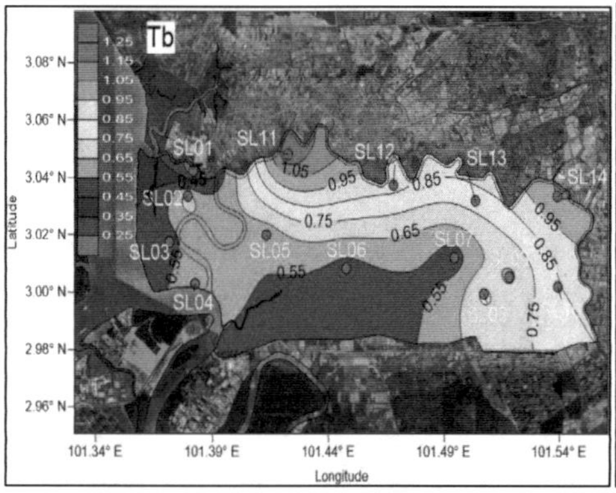

(o)

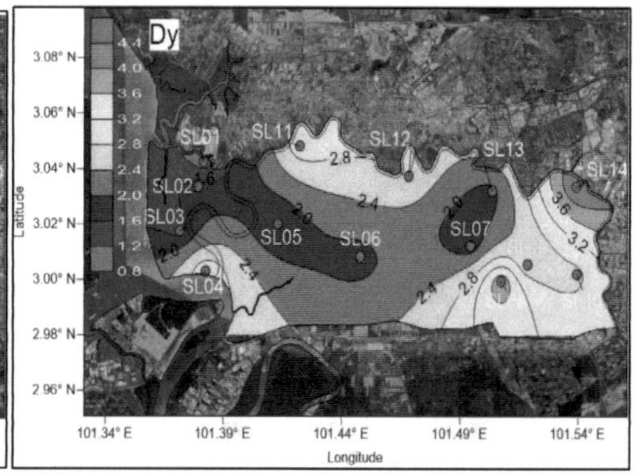

(p)

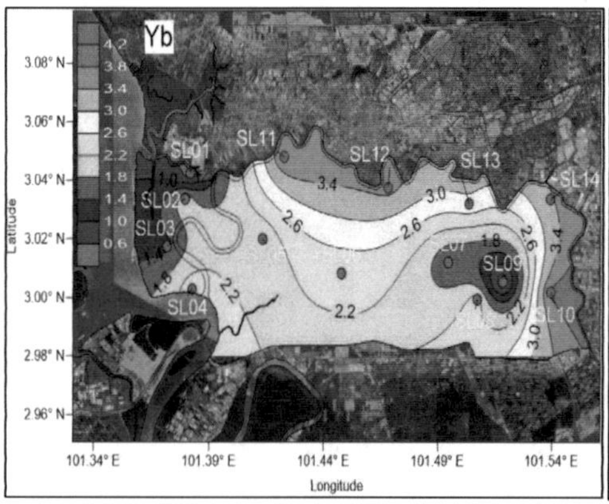

(q)

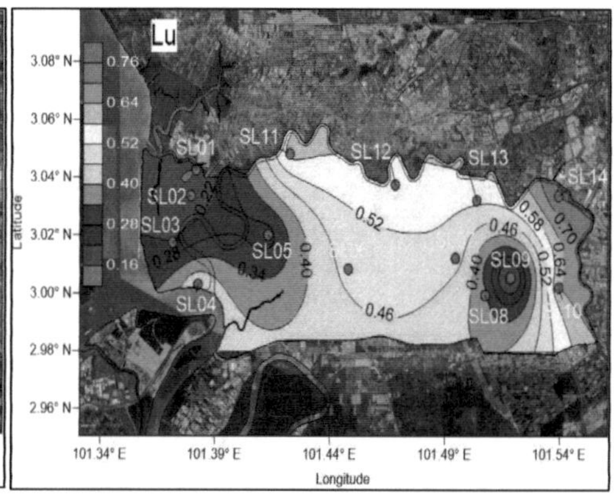

(r)

Figure 3 (a) – (r). Contour map of major elements (Fe, Mg, Ti), heavy metals (As, Cr, Sb, Zn) actinide (U, Th) and rare earth elements (La, Ce, Nd, Sm, Eu, Tb, Dy, Yb, Lu) concentration distribution in the soil of Klang industrial area.

4. Conclusion

The major elements of Mg and Ti showed higher concentrations in the SL14 location, which is located a food and beverage manufacturers, rubber, recycling centre, paper, chemical, and plastic industries. The arsenic (As) showed a higher concentration in the SL11 location, which is located a residential, shop lots and vehicle workshops. The possible source of higher concentration of arsenic is due to release or seepage of lubricant oil waste from the motor vehicle workshop. Zn concentration showed more elevated in the SL07 and SL08 locations. The main activities in that area are workshops, concrete, manufacturing, plastic, motor, and machine industries. The Cr and Sb showed higher concentrations in the SL14 location. The geo-accumulation (I_{geo}) index of major elements can be categorised as uncontaminated to moderately contaminated. As and Sb can be categorised as uncontaminated to moderately contaminated and as extremely contaminated. The I_{geo} index of Cr and Zn can be categorised as uncontaminated to strongly contaminated. REEs such as La, Ce, Nd, Sm, Eu, Tb, Dy, Yb, and Lu showed higher in the SL 14 than other areas, which are located food & beverage manufacturers, rubber, recycling centre, paper, chemical, plastic industries. The I_{geo} index values of REEs can be categorised as uncontaminated in the soil of the Klang Industrial area. The U and Th concentrations showed higher in the SL09 and SL10, which are chemical, electronic, paint, recycling paper, plastic, and rubber industries. The I_{geo} index of U and Th can be categorised as uncontaminated and as moderately and strongly contaminated.

5. References

[1] Elias M S, Azman M A, Daung J A D, Hashim A, Laili Z, Omar S A and Shukor S 2022 Assessment of rare earth and actinides (U and Th) elements in soil samples from Kapar industrial area, Selangor *IOP Conf. Ser.: Mater. Sci.and Eng.* **1231** 012017.

[2] Yap C K, Edward F B and Tan S G 2010 Heavy metal conterations (Cu, Pb, Ni and Zn in the surface sediments from a semi-enclosed intertidal water, the Johore Straits: Monitoring data for future reference *J. of Sustain. and Manag.* **5** 44–57.

[3] Ashraf A, Saion E, Gharibshahi E, Kamari H M, Yap C K, Hamzah M S and Elias M S 2017 Distribution of Trace elements in core marine sediments of coastal east Malaysia by instrumental neutron activation analysis *Appl. Radiat. and Isot.* **122** 96–105.

[4] Elias M S, Ibrahim S, Samuding K, Daung J A D, Kantasamy N, Rahman S A and Hashim A 2020 Distribution, sources and modified degree of contamination of trace elements in Linggi estuary surface sediments, Malaysia *J. of Sustain. and Manag.* **15** 109–19.

[5] Elias M S, Wo Y M, Hamzah M S, Shukor S A, Rahman S A, Salim N A A, Azman M A and Hashim A 2017 Analysis of soil samples from gebeng area using NAA technique *AIP Conf. Proc.* **1799** 030008-1-030008–11.

[6] Siti Norbaya Mat Ripin, Sharizal Hasan, Mohd Lias Kamal N M H 2014 Analysis and pollution assessment of heavy metal in soil , Perlis *The Malaysian J.of Analy. Sc.* **18** 155–61.

[7] Kamarudzaman A N, Woo Y S and Ab Jalil M F 2018 Distribution and analysis of heavy metals contamination in soil, Perlis, Malaysia *E3S Web of Conf.* **34** 1–5.

[8] Azman M A, Hamzah S, Rahman S A, Elias M S, Abdullah N, Hashim A, Shukor S A and Kamaruddin A H C 2015 Trace element analysis of soil type collected from the manjung and central Perak *AIP Conf. Proc.* **1659** 050003–1 – 050003–16.

[9] Aysha Masood K 2017 Speciation and bioavailibility of rare earth elements (rees) in ex-tin mining area: a case study from Lahat, Perak, Malaysia PhD Thesis (University of Malaya) **1** 1–243.

[10] Elias M S, Ibrahim S, Samuding K, Rahman A R and Hashim A 2018 The sources and ecological risk assessment of elemental pollution in sediment of Linggi estuary , Malaysia *Marine Pollut. Bull.* **137** 646–55.

[11] Joel E S, Maxwell O, Adewoyin O O, Ehi-Eromosele C O, Embong Z and Oyawoye F 2018 Assessment of natural radioactivity in various commercial tiles used for building purposes in Nigeria *MethodsX* **5** 8–19

[12] Rezaee K, Saion E B, Wood A K and Abdi M R 2010 Rare earth elements determination and distribution patterns in surface marine sediments of the South China Sea by INAA *J. of Radioanaly. and Nucl. Chem.* **283** 823–9.

[13] Turekian K K and Wedepohl K H 1961 Distribution of the elements in some major units of the Earth's crust *Geol. Soc. Am. Bull* **72** 175–92.

[14] Abrahim G M S and Parker R J 2008 Assessment of heavy metal enrichment factors and the degree of contamination in marine sediments from Tamaki Estuary, Auckland,

New Zealand *Environ. Monit. and Assess.* **136** 227–38.

[15] Elias M ., Mohd Suhaimi H, Shamsiah A R, Wee B S and Nazaratul A A S 2012 Assessment of Sediment quality collected from the Tunku Abdul Rahman National Park , Sabah *Jurnal Sains Nuklear Malaysia* **24** 59–70.

[16] Nobi E P, Dilipan E, Thangaradjou T, Sivakumar K and Kannan L 2010 Geochemical and geo-statistical assessment of heavy metal concentration in the sediments of different coastal ecosystems of Andaman Islands, India *Estuar. Coastal and Shelf Sci.* **87** 253–64.

[17] Muller G 1969 Index of geoaccumulation in sediments of the Rhine river *Geojour.* **2** 108–18.

[18] Yin H, Gao Y and Fan C 2011 Distribution, sources and ecological risk assessment of heavy metals in surface sediments from Lake Taihu, China *Environ. Res. Lett.* **6** 1–11.

[19] Zhu L, Xu J, Wang F and Lee B 2011 An assessment of selected heavy metal contamination in the surface sediments from the South China Sea before 1998 *J. of Geochem. Explorat.* **108** 1–14

[20] Elias M S 2021 Distribution and source identification of heavy metals, trace and rare earth elements in surface sediment of Linggi River PhD Thesis (Universiti Teknologi MARA) 1–309.

Acknowledgement

The authors would like to thank the Malaysian Nuclear Agency and the Regional Cooperative Agreement Regional Office (RCARO) for financial support under the research grant (RCARP02/RC03). We would also like to express our gratitude and special thanks to the Analytical Chemistry Application Group staff for their sampling and analysis of samples.

The potential of NAA and ICP-MS for determining soil dispersibility in relation to landslide incidents

A S N Shah[1,2,a], M A Zali[3], N S M Nazer[2], A Hashim[3], A Rusdi[3] and A H Ismail[4]

[1]Materials Technology Group, Industrial Technology Division, Malaysian Nuclear Agency, 43000 Kajang, Selangor, Malaysia
[2]Department of Earth Sciences and Environment, Faculty of Science and Technology, National University of Malaysia, 43600 Bangi, Selangor, Malaysia
[3]Analytical Chemistry Application Group, Waste Technology & Environmental Division, Malaysian Nuclear Agency, 43000 Kajang, Selangor, Malaysia
[4]Plant Assessment Technology Group, Industrial Technology Division, Malaysian Nuclear Agency, 43000 Kajang, Selangor, Malaysia

[a]azlanshah@nm.gov.my

Abstract. Dispersive soils pose significant challenges due to their inherent tendency to rapidly respond to the presence of water, resulting in erosion and heightened susceptibility to landslides. The inclusion of sodium (Na) onto the surface of soil clay particles results in the attenuation of the electrochemical interactions between the dispersed soil, leading to the repulsion and subsequent migration of the clay particles away from one another. This study employed Neutron Activation Analysis (NAA) and Inductively Coupled Plasma Mass Spectrometry (ICP-MS) to assess the Sodium Adsorption Ratio (SAR) and Exchangeable Sodium Percentage (ESP) as indicators of soil dispersibility. The results indicate that soils containing montmorillonite clay mineral exhibit greater values of SAR > 11 and ESP > 20 in comparison to soils containing other clay minerals i.e. kaolinite, illite, and vermiculite. The calculated SAR and ESP values obtained from both analyses demonstrate a notable agreement, indicating a positive correlation. In conclusion, it is apparent that both NAA and ICP-MS method has the capability to be applied for the purpose of predicting soil dispersibility in landslide-prone regions of Malaysia.

1. Introduction

Dispersive soil or known as sodic soil is a soil type that poses challenges due to its propensity to readily interact with water, primarily attributed to the presence of sodium ions (Na^+) generated by mineralogy of the parent material, topography, and water table quality [1]. While it is commonly believed that the occurrence of dispersive soil is limited to region with arid or semi-arid climates, other countries globally have encountered challenges associated with this phenomenon [2, 3]. Dispersive soil has been reported in various tropical nations, including Thailand, Vietnam, Indonesia, and Malaysia [4, 5, 6, 7]. Several regions of Malaysia, prior studies have documented the presence of this particular type of soil in Batu Gajah, Wangsa Maju, Shah Alam and East Sabah. These places are distinguished by the prevalence of high clay content and clay particles [6, 8, 9]. The susceptibility of dispersive soil to develop erosion renders it a contributing factor to heightened slope instability. This erosion can manifest in the form of internal erosion or pipe erosion, which is often triggered by soil cracking [10]. Consequently, the

IOP Conf. Series: Materials Science and Engineering 1308 (2024) 012018 doi:10.1088/1757-899X/1308/1/012018

occurrence of landslides is exacerbated [11]. The desiccation process leads to the development of crusts and fracture networks, which subsequently diminish the pore area on the soil surface. This reduction in pore area has significant implications for several soil properties and processes, including infiltration rates, surface water runoff, bearing capacity, and permeability. Previous research has indicated that the dispersion of montmorillonite minerals in soil is influenced by the presence of sodium (Na) derived from mineral sources rich in Na-plagioclase, such as albite [12]. The presence of Na-montmorillonite in dispersive soil leads to the development of a second electrical layer on its surface. This phenomenon has the effect of reducing the bonding strength between mineral layers and the interparticle interactions within the lamellar structure [13].

Dispersive clay soils constitute a distinct group of fine-grained soils that provide challenges in their identification using visual categorization or physical examination techniques, such as consistency tests and particle size analysis [14]. The crumb and double hydrometer tests are well recognized as straightforward and preliminary techniques for evaluating the dispersive characteristics of soil. The pinhole test provides distinct results for the categorization of soils into dispersive and non-dispersive groups, which is determined by analyzing the effluent and diameter of the drilled specimen. However, the results of the tests resulted in a certain degree of uncertainty in the categorization procedure, primarily due to the presence of sand-clay particles percentage. In instances of this nature, it becomes imperative to employ chemical testing to validate the dispersion characteristics within the soil. Therefore, the assessment of soil dispersibility can be accomplished through the application of Soil Exchangeable Sodium Percentage (ESP) and Sodium Adsorption Ratio (SAR). Soils can be characterized as dispersive when the SAR > 13 and ESP > 15 [15]. To provide more clarification, the objective of this research endeavor is to assess the viability of utilizing NAA (Neutron Activation Analysis) and ICP-MS (Inductively Coupled Plasma Mass Spectrometry) techniques for the purpose of ascertaining the dispersibility of soil by estimating the SAR and ESP values. Both analytical techniques are commonly employed to quantify the constituents in diverse applications. NAA was seen as a comparatively more convenient technique for handling soil samples in comparison to ICP-MS, as the latter entails the utilization of potent acids during the sample preparation phase. NAA enables the concurrent and precise identification of several elements with a minimal quantity of solid matrix, in contrast to ICP-MS, which is better suited for analyzing liquid samples (dissolved solid) [16, 17].

2. Material and methodology

The process flow in the present study is depicted in Figure 1. A collection of 10 soil samples with disturbed clayey composition was obtained from locations that experienced landslides. The selection of these samples was based on the relative abundance of four primary clay minerals identified in the soil, namely kaolinite (K), illite (I), montmorillonite (smectite) (M), and vermiculite (V), as determined using mineral phase characterization. According to the results of physical tests, specifically the crumb and pinhole dispersion test, the montmorillonite soil samples were categorized as dispersive soil, indicating that they exhibited a tendency to disperse when in contact with water. In contrast, the other soil samples were classed as non-dispersive soil, as they did not exhibit any reaction when exposed to water. The soil samples underwent a process of oven-drying at a temperature of 105°C for a duration of 24 hours. Following this, the samples were crushed and then sieved. The subsequent step involved the preparation of the samples for NAA and ICP-MS in order to quantify the elemental composition and ascertain the SAR and ESP, thereby reaffirming the observed dispersibility phenomena.

2.1. Neutron activation analysis (NAA)

NAA is a very sensitive analytical technique that enables the reliable and precise determination of constituent concentrations in soil samples. The soil samples, each weighing 1 g, were prepared and thereafter placed in polyethylene tubes prior to the irradiation process. The specimens, along with standard reference materials (SRMs) with predetermined elemental concentrations, are subjected to irradiation by thermal neutrons within a nuclear reactor. The irradiation of all samples and SRMs was performed in the 500 kW PUSPATI TRIGA Mark II research reactor at the Malaysian Nuclear Agency.

IOP Conf. Series: Materials Science and Engineering 1308 (2024) 012018 doi:10.1088/1757-899X/1308/1/012018

The reactor provided a thermal neutron flux of 4.0×10^{12} n cm^{-2} s^{-1}. Following a suitable decay time, the utilization of high-resolution gamma ray spectroscopy is employed to quantify the intensity and energy of gamma lines generated by radioactive isotopes. A comparative analysis was conducted between the induced activity in the SRMs and the sample in order to ascertain the elemental abundances. The concentrations of sodium (Na), magnesium (Mg), and calcium (Ca) were measured in milliequivalents per litre (meq/L).

2.2. Inductively coupled plasma mass spectrometry (ICP-MS)
Soil samples to be analyzed by ICP-MS require acid digestion to ensure element dissolution and extraction. This study used 0.5 g of dry soil powder that was dissolved in 2 mL hydrochloric acid, 3 mL nitric acid and 5 mL hydrofluoric acid in a vessel and allowed to react for 15 minutes before sealing it. The vessels were transferred to a vessel liner and ramp at 175°C for 15 minutes before holding for another 15 minutes and thereafter cooling them for 30 minutes. The digested content was then emptied to a plastic container, labeled and stored at 4°C for analysis purpose.

2.3. Sodium adsorption ratio (SAR) and exchangeable sodium percentage (ESP)
The SAR and ESP play a crucial role in assessing the existence of soil dispersion resulting from the presence of exchangeable sodium. The calculation of the SAR was performed using the measured concentrations of sodium (Na), calcium (Ca), and magnesium (Mg) in milliequivalents per millilitre (meq/L), as described by Figure 1. The SAR does not incorporate monovalent potassium (K) and assumes an equal treatment of magnesium (Mg) and calcium (Ca) [17]. ESP was estimated by calculating the ratio of exchangeable sodium (Na) to the cation exchange capacity (CEC) measured in milliequivalents per 100 grams (meq/100g) using the methylene blue spot method [18].

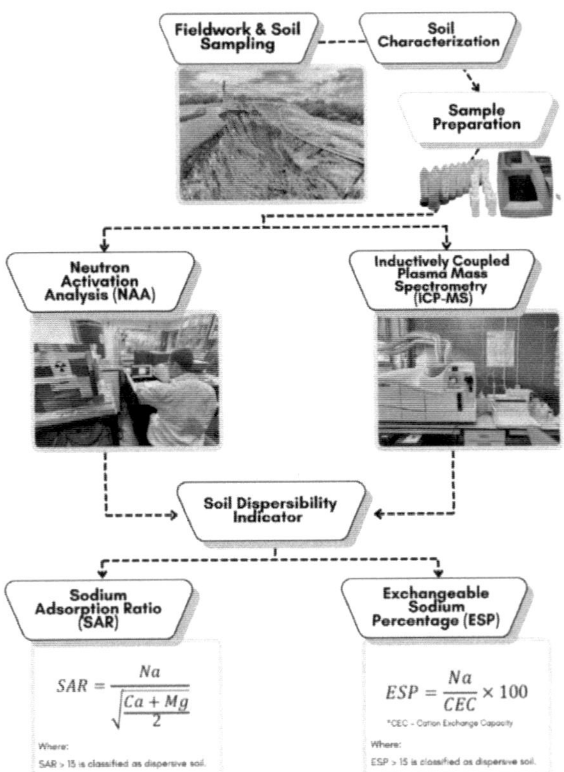

Figure 1. Flow of methods in this study.

3. Results and discussion

Table 1 displays the concentration of Na, Mg and Ca measured by NAA and ICP-MS. From the results, NAA detect Na and Mg for all samples except for Ca while ICP-MS is less sensitive to Na compared to Mg and Ca elements. This is probably by some elements in soil particles were not entirely dissolved by the acid mixtures. The results of SAR and ESP calculated from NAA and ICP-MS elemental quantification are presented in Table 2 and Figure 2. Montmorillonite soil samples mineral exhibit greater values of SAR > 11 and ESP > 20 in comparison to soils containing other clay minerals i.e. kaolinite, illite, and vermiculite. The results obtained from the standard test provide further confirmation that the soil exhibits increased dispersibility as a result of the presence of Na-plagioclase rich minerals. Results indicated a change in the SAR and ESP in the following order: montmorillonite, illite, kaolinite, vermiculite. These expanding clay minerals have a greater tendency to disperse compared to non-expanding clay minerals. On the other hand, vermiculite exhibits a lower sensitivity to dispersion as a result of its relatively low Na content compared to other soluble salts (Mg and Ca) in the soil. Low Na concentrations increase the attractive Van der Waals interactions between soil particles and decrease osmosis potential, minimising soil erosion.

Table 1. Concentration of elements and CEC values.

Soil Sample No.	NAA (meq/L)			ICP-MS (meq/L)			CEC (meq/100g)
	Na	Mg	Ca	Na	Mg	Ca	
K1	7.6	N.D.	N.D.	N.D.	179.2	23.7	21.7
K2	3.7	298.9	N.D.	N.D.	53.4	5.8	8.7
K3	2.7	382.9	N.D.	N.D.	87.3	20.2	26.0
I1	40.6	2627.1	N.D.	8.9	101.3	3.5	30.4
I2	10.1	3011.9	N.D.	N.D.	53.4	5.8	21.7
I3	8.0	1485.9	N.D.	N.D.	87.3	20.2	21.7
M1	424.1	963.0	N.D.	342.7	418.1	293.4	130.2
M2	412.9	881.5	N.D.	266.6	302.8	242.2	125.9
M3	264.2	786.9	N.D.	201.5	450.4	131.4	117.2
V1	19.3	10168.0	N.D.	20.8	4003.2	202.7	134.6

Table 2. Value of SAR and ESP calculated from elemental detection by NAA and ICP-MS

Soil Sample No.	NAA			ICP-MS		
	SAR	ESP (%)	Dispersibility	SAR	ESP (%)	Dispersibility
K1	-	3.5	Non-dispersive	-	-	Non-dispersive
K2	0.30	4.3	Non-dispersive	-	-	Non-dispersive
K3	0.3	3.7	Non-dispersive	2.9	6.6	Non-dispersive
I1	1.1	13.4	Non-dispersive	1.4	3.8	Non-dispersive
I2	0.3	4.7	Non-dispersive	-	-	Non-dispersive
I3	0.3	3.7	Non-dispersive	-	-	Non-dispersive
M1	19.3	32.6	Dispersive	18.1	26.3	Dispersive
M2	19.7	32.8	Dispersive	16.1	21.1	Dispersive
M3	13.3	22.5	Dispersive	11.8	17.2	Moderately dispersive
V1	0.3	1.4	Non-dispersive	0.5	1.5	Non-dispersive

Note: K – Kaolinitic soil; **I** – Illite soil; **M** – Montmorillonite soil; **V** – Vermiculite soil

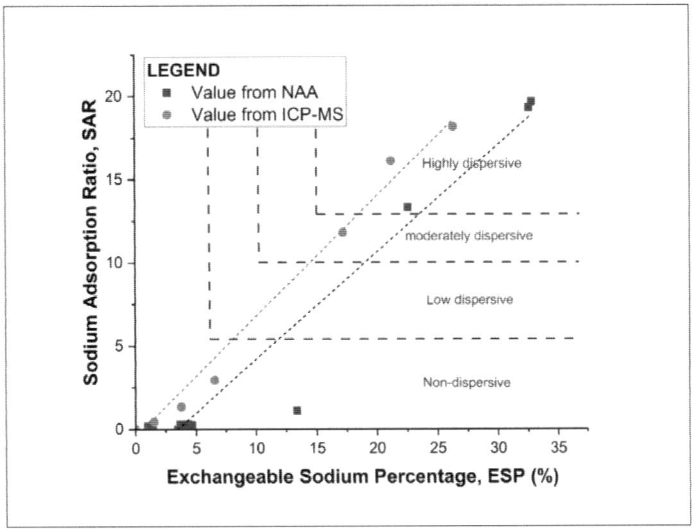

Figure 2. Relationship between ESP and SAR value from NAA and ICP-MS to soil dispersibility, respectively.

Figure 3 shows the correlation of SAR and ESP value determined by NAA and ICP-MS. From the graph, the pearson correlation matrix value for SAR and ESP of both analytical methods are r = 0.9897; R^2 = 0.9795 and r = 0.93917; R^2 = 0.88205, respectively. The value showing a good agreement, whereas most of samples fall in the same categories (non-dispersive to highly dispersive). This is noteworthy considering the vast range of element concentrations observed within each respective method.

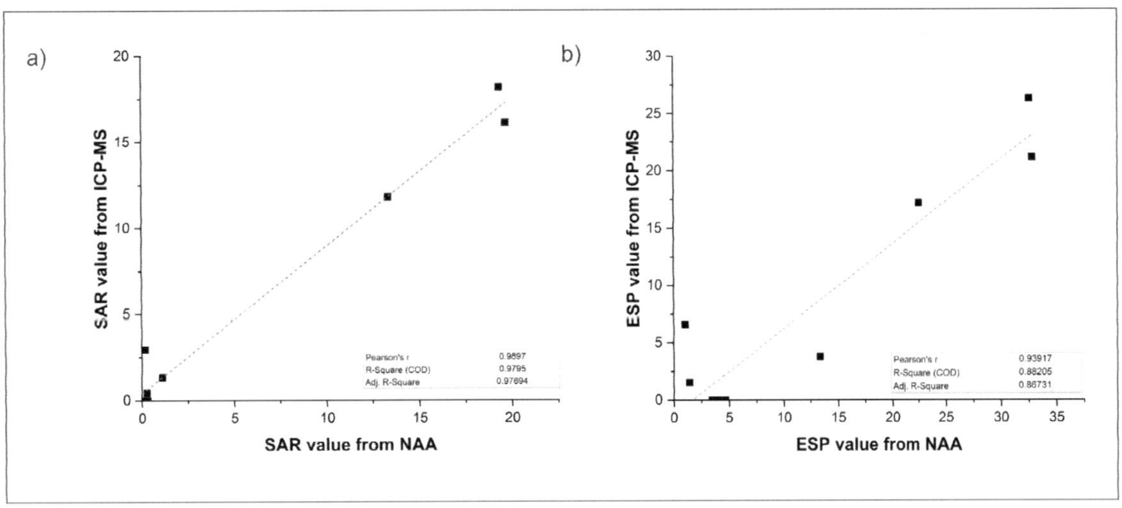

Figure 3. The figure illustrated: a) Correlation between SAR value from NAA and ICP-MS analysis; and b) Correlation between ESP value from NAA and ICP-MS analysis.

4. Conclusion

In conclusion, the utilization of NAA and ICP-MS can be employed as analytical techniques to assess the soil dispersibility caused by the presence of Na in soil. The calculated SAR and ESP values obtained from both analyses demonstrate a good agreement, indicating a positive correlation. The observed variations in element concentrations throughout the samples can be attributed to a variety of factors, including sample preparation techniques and instrumental limitations such as partial sample dissolution and detection limits. Soils that are predominantly composed of montmorillonite clay minerals typically exhibit higher values SAR > 11 and ESP > 20, when compared to soils containing other clay minerals such as kaolinite, illite, and vermiculite. This can be attributed to the presence of Na-plagioclase rich minerals in the soil. In addition to topography and hydrology, the onset of landslides is also impacted by soil factors, particularly the presence of Na in the soil, which enhances the erodibility of slopes.

5. References

[1] Stavi I, Thevs N, and Priori S 2021 Soil Salinity and Sodicity in Drylands: A Review of Causes, Effects, Monitoring, and Restoration Measures *Front. Environ. Sci.* **9** 712831 doi:10.3389/fenvs.2021.712831.

[2] Hassanlourad M, Rokni M N, Hassanlo M, and Badrlou, A 2017 Dispersive clay stabilised by alum and lime *Int. J. Geomate* **12** 156-162.

[3] Premkumar S, Piratheepan J, Arulrajah A, Disfani M M, and Rajeev P 2016 Experimental study on contact erosion failure in pavement embankment with dispersive clay *J. Mater. Civ. Eng.* **28** 04015179.

[4] Munirwan R P, and Munirwansyah, M 2022 Assessing slope failure of soil erodibility problem by soil dispersive identification *E3S Web Conf.* **340** 01006.

[5] Patcharapreecha P, Topark Ngarm B, Goto I, and Kimura M 1990 Studies on saline soils in Khon Kaen Region, Northeast Thailand *Soil Sci. Plant Nutr.* **36**(3) 363-374.

[6] Shah A S N, Nazer N S M, Harris M I 2022 Morfologi Hakisan dan Sifat Serakan Lempung Kaolinit dan Montmorilonit di Kawasan Tropika *Sains Malays.* **51**(12) 3879-3896 http://doi.org/10.17576/jsm-2022-5112-02.

[7] Tran D B, Hoang T V, and Dargusch P 2015 An assessment of the carbon stocks and sodicity tolerance of disturbed Melaleuca forests in Southern Vietnam *Carbon Balance Manag.* **10(15)** https://doi.org/10.1186/s13021-015-0025-6.

[8] Ismail F, Mohamed Z, and Mukri M 2008 A study on the mechanism of internal erosion resistance to soil slope instability *Electron. J. Geotech. Eng.* **13(A)** 1-12.

[9] Vakili A H, bin Selamat M R, Aziz H B A, Mojiri A, Ahmad, and Safarzadeh M 2017 Treatment of dispersive clay soil by ZELIAC *Geoderma* **285** 270-279.

[10] Vakili A H, Shojaei S I, Salimi M, bin Selamat, M R and Farhadi M S 2020 Contact erosional behaviour of foundation of pavement embankment constructed with nanosilica-treated dispersive soils *Soils Found.* **60(1***)* 167-178 https://doi.org/10.1016/j.sandf.2020.02.001.

[11] Wang L, Yuan X and Wang M 2020 Landslide failure mechanisms of dispersive soil slopes in seasonally frozen regions *Adv. Civ. Eng.* **2020** 1-13.

[12] Shah A S N and Zali M A 2023 Potential of tropical expansive clays as natural sealing layer in nuclear waste repository system: a preliminary study *IOP Conf. Ser.: Mater. Sci. Eng.* **1285** 012018.

[13] Liu J, Chen P, Lu Z and Yao H 2022 Experimental study on the modification mechanisms of dispersive soil treated with hydroxyl aluminum *Geofluids* **2022** 1–13.

[14] Djoković K, Čaki L, Šušić N and Hadži-Niković G 2018 Methods for assessment and identification of dispersive soils *Ce/papers* **2(2-3)** 205–210 https://doi.org/10.1002/cepa.672.

[15] Abbaslou H, Hadifard H and Ghanizadeh A R 2020 Effect of cations and anions on flocculation of dispersive clayey soils *Heliyon* **6(2)** e03462.

[16] Moens L and Dams R 1995 NAA and ICP-MS: A comparison between two methods for trace and ultra-trace element analysis *J. Radioanal. Nucl. Chem.* **192**(1) 29–38.

[17] Tsukada H, Takeda A, Hasegawa H, Ueda S and Iyogi T 2005 Comparison of NAA and ICP-MS for the determination of major and trace elements in environmental sample *J. Radioanal. Nucl. Chem.* **263**(3) 773–778.
[18] Rengasamy P, Tavakkoli E and McDonald G K 2016 Exchangeable cations and clay dispersion: net dispersive charge, a new concept for dispersive soil. *Eur. J. Soil Sci.* **67(5)** 659-665.
[19] Yukselen Y and Kaya A 2008 Suitability of the methylene blue test for surface area, cation exchange capacity and swell potential determination of clayey soils *Eng. Geol.* **102** 38-45. http://dx.doi.org/10.1016/j.enggeo.2008.07.002. 3

Acknowledgements
The authors would like to thank the Materials Technology Group (MTEG) and Analytical Chemistry Application Group (ACA) of the Malaysian Nuclear Agency for supporting this research. This gratitude also extends to the Department of Earth Sciences and Environment at Universiti Kebangsaan Malaysia for their guidance and assistance.

IOP Conf. Series: Materials Science and Engineering 1308 (2024) 012019 doi:10.1088/1757-899X/1308/1/012019

A study on surface air Beryllium-7 concentration at radionuclide station in Tanah Rata, Cameron Highlands (MYP42) during southwest monsoon seasons in Malaysia

Mohd Fauzi Haris[1,2,a)], Norita Md. Norwawi[1,b)], Mohd Hafez Mohd Isa[3,c)], Muhammad Rawi Mohamed Zin[2,d)], Muhammed Zulfakar Zolkaffly[2,e)], Azlai Ta'at[4,f)]

[1]Cybersecurity and Systems Research Unit, Faculty of Science and Technology, Universiti Sains Islam Malaysia, 71800 Nilai, Negeri Sembilan, Malaysia
[2]Malaysian Nuclear Agency, 43000 Kajang, Selangor, Malaysia
[3]Faculty of Science and Technology, Universiti Sains Islam Malaysia, 71800 Nilai, Negeri Sembilan, Malaysia
[4]Malaysian Meteorological Department, Jalan Sultan, 46667 Petaling Jaya, Selangor, Malaysia

[b]norita@usim.edu.my,
[a]fauziharis@nm.gov.my
[c]m.hafez@usim.edu.my
[d]muhammad_rawi@nm.gov.my
[e]zulfakar@nm.gov.my
[f]azlai@met.gov.my

Abstract. Beryllium-7 (Be-7) is a significant radionuclide, primarily produced through cosm ray interactions with atmospheric elements. Its study can provide insights into atmosphe ic processes and transport mechanisms. This research investigates the concentration of Be-7 ric surface air at the Radionuclide Station (MYP42) located in Tanah Rata, Cameron Highlands n Malaysia, during the Southwest Monsoon season using data from the Comprehensive Nuclea , Test-Ban Treaty Organisation (CTBTO) from 2011 to 2021. Meteorological parameters of t r- study area, such as wind direction, wind speed, humidity, and precipitation, are considered the understand their influence on radionuclide dispersion and deposition. The results from this stu o are aimed to contribute to the understanding of atmospheric dynamics in the Cameron Highlands dy region and may have implications for broader atmospheric and environmental studies. The stu results suggest no statistically significant variation in Be-7 concentration during the southwe dy monsoon season, demonstrating minimal fluctuations within and outside the monsoon period st Such insights are crucial for the region's scientific research and environmental management. .

1 Introduction

The meteorological phenomena known as the Southwest Monsoon (SWM), from the end of May t mid-September, and the Northeast Monsoon (NEM), which occurs from November to March, constitu o tw o distinct monsoonal systems exerting a profound influence upon the climatic patterns observed te Ma laysia. The monsoon transition period is the name given to the period of interstitial space betwe n the se two monsoonal systems. This study, however, focuses on Be-7 concentration during the SWM en , which tends to be comparatively less humid across the nation, except for the state of Sabah in East Malaysia.

Malaysia has precipitation throughout the year. Annually, The Southwest Monsoon (SWM) season typically spans from late May to September. Throughout this period, the entire country encountered a drier atmosphere, especially along the western coast of the Malaysian Peninsula [1]. During this season, the predominant wind originates from the southwest and is associated with reduced air humidity and a more stable atmospheric state. This monsoon is characterized by lower precipitation, diminished cloud cover, higher outgoing long-wave radiation (OLR), and frequent dry spells [2]. Due to the lower rainfall, particularly in Peninsular Malaysia's northwest, it is often frequently referred to as the summer season [3]. In addition, most regions have experienced a decrease in rainfall during the Southwest Monsoon during the past three decades [4]. As summer progresses, open burning poses a greater threat to the environment and produces significant haze, which may pose serious health risks. [1].

In a different vein, Be-7, a naturally occurring cosmogenic radionuclide, is formed through continuous spallation reactions in the upper troposphere and lower stratosphere, as explained by [5]. Once created, it promptly associates with minuscule particles. The atmospheric presence of Be-7 is uneven, influenced by geographical location, solar activity, the interchange of air masses within the Atmosphere, and its removal efficiency from the troposphere [6]. With an energy of 477.6 keV and a half-life of 53.3 days, which is relatively short [7], it is easily detected and identified using gamma-spectrometry with an HPGe detector. Due to its convenient half-life and easily detectable radiation, Be-7 holds great significance as an isotope for investigating atmospheric processes and weather-related occurrences. It has found applications in the examination of precipitation scavenging, the removal of air masses in both vertical and horizontal directions, the transit and residence times of aerosols in the troposphere, aerosol deposition rates, and the deposition patterns of airborne pollutants, as outlined by [8]. The study by [9] highlights the notable influence of precipitation in bringing about seasonal variations in fine aerosol levels, especially during the summer monsoon phase in Northeast Asia. Their study unveils distinctive patterns in Be-7 concentrations, particularly highlighting the significant finding of a negative or inverse correlation between Be-7 levels in surface air and precipitation. Inverse correlation refers to a relationship between two variables moving in opposite directions. When one variable increases, the other decreases, and vice versa. Furthermore, they emphasized the existing gaps in our understanding of the seasonal variations in aerosol removal concerning precipitation, suggesting the need for further research to better comprehend the connection between Be-7 and fine aerosols in the future.

The study by [10] documented a correlation between various meteorological factors and Be-7 levels in Tanzania. Their findings revealed a noteworthy negative correlation with rainfall and absolute humidity and a relatively modest positive correlation with wind speed. Such a correlation pattern was also revealed by a study reported in [11] while studying Be-7 in Northern Europe. In addition to this study, previous research has extensively explored the connections between meteorological parameters and Be-7 levels, referencing the discoveries outlined in the studies of [12], [13], [14], [15], and [16]. As explored in the study by [17], relative humidity emerges as one of the pivotal determinants in the seasonal fluctuations of Be-7. Evaluating the dispersion of natural radionuclides in the Atmosphere offers a valuable means of examining air mass circulation. A more in-depth examination of meteorological effects reveals that the impact of meteorological factors on Be-7 concentrations varies based on geography and season [18].

1.1. Study objectives
The primary goal of this study is to examine the levels of Be-7 concentration observed in the surface air of Tanah Rata, Cameron Highlands, Malaysia, spanning from the SWM season of 2011 to 2021. Additionally, the research aims to investigate the relationship between these Be-7 concentrations and the distinct meteorological elements in the study region. Furthermore, it also aims to discover how atmospheric conditions and processes impact airborne radioactivity levels during the SWM seasons.

2. Materials and methodology

2.1. Location of study

The CTBTO Radionuclide Monitoring Station (MYP42), located at 4.48°N and 101.37°E in Tanah Rata, Cameron Highlands, is depicted in Figure 1. This station, situated at an elevation of 1545 meters above sea level, is a facility for monitoring particulate radionuclides and shares its location with the Malaysian Meteorological Station in Tanah Rata, Cameron Highlands, Pahang, as detailed in the study by [19]. The facility is part of the Comprehensive Nuclear-Test-Ban Treaty Organization (CTBTO) International Monitoring System (IMS) and is one of the 80 radionuclide monitoring stations responsible for tracking atmospheric radionuclide particles that signal a nuclear detonation.

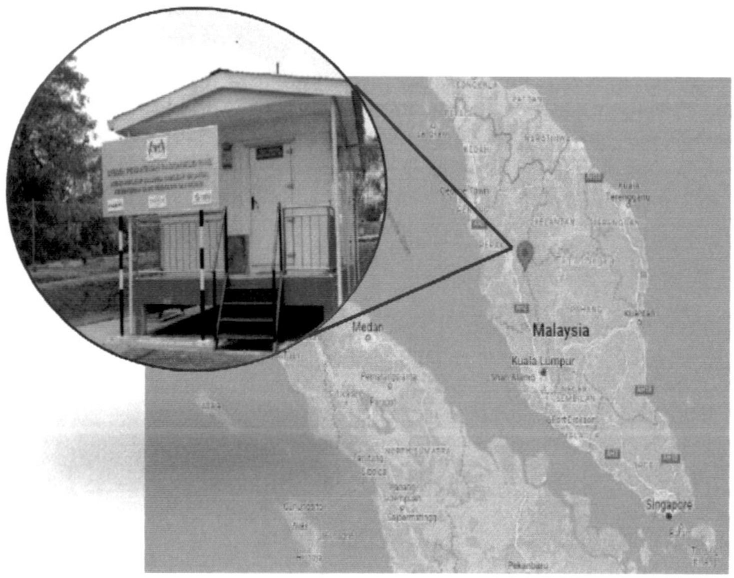

Figure 1. The location of MYP42 station.

2.2. Be-7 intensity concentration database

In this investigation, we collected the daily activity concentrations of Be-7 captured by the MYP42 station from 2011 to 2021. These records were sourced through the CTBTO International Data Centre (IDC) using the RNToolkit version 0.3.5. Figure 2 illustrates the RNToolkit interface. Due to the tool's limitation in displaying data for an extended period, we split the request into figures: Figure 2(a) and Figure 2(b), showcasing the Be-7 concentration at MYP42 for 2011-2015 and 2016-2021, respectively. Additionally, the IDC serves as a repository for databases related to three other global monitoring technologies designed to detect potential nuclear detonations: seismic, hydroacoustic, and infrasound (SHI).

IOP Conf. Series: Materials Science and Engineering 1308 (2024) 012019 doi:10.1088/1757-899X/1308/1/012019

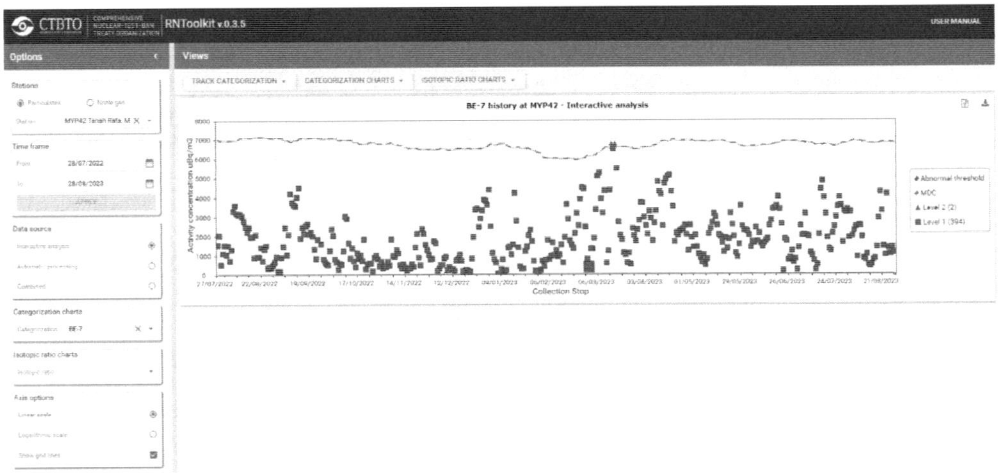

Figure 2. RNtoolkit interface with its options on the left-hand side.

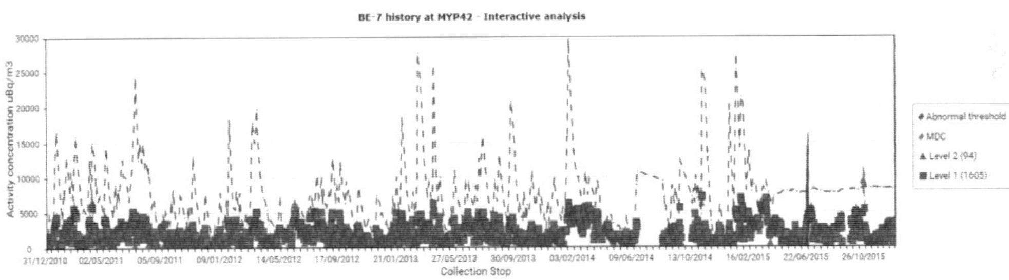

Figure 2(a). Be-7 intensity concentration for 2011-2015.

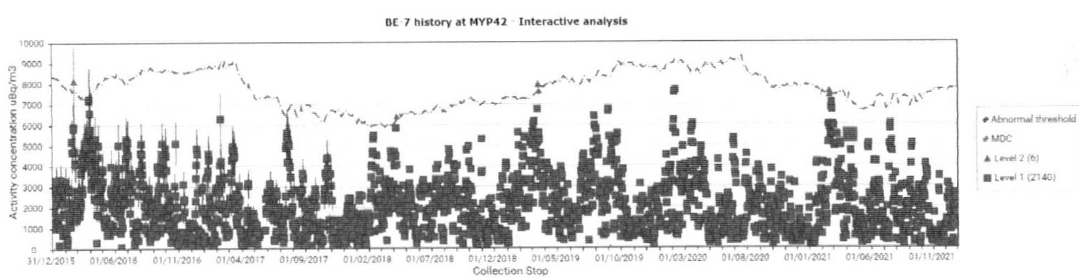

Figure 2(b). Be-7 intensity concentration for 2016-2021.

2.3. Meteorological variables

The daily records of meteorological information in this study, encompassing relative humidity, air pressure, wind speed, wind direction, and precipitation, were acquired from the Malaysian Meteorological Department (MET Malaysia).

2.4. Statistical analysis

A Two-sample t-test is a statistical hypothesis test employed to assess if a noteworthy distinction exists between the means of two independent groups. In this context, it is employed to compare the mean values of Be-7 recorded for both the SWM and non-SWM periods to ascertain whether statistically significant differences exist between those datasets. Assuming that the variance of these two groups of data is different, Welch's t-test is used.

Conversely, the correlation coefficient is a statistical metric that quantifies the strength and direction of the association between variables. It is applied in this study to evaluate the relationship between Be-7 and specific meteorological parameters. The outcomes of the correlation analysis are subsequently applied in a Fisher Z-transformation, which facilitates the comparison of correlations between the two groups of interest, the SWM and Non-SWM. This procedure helps determine if the differences in correlations carry statistical significance. The interpretation of the correlation value is used in Table 1 below.

Table 1. Correlation coefficient ranges and interpretations [20].

Coefficient $\|r\|$	*Interpretation*
0.00 – 0.09	No or negligible
0.10 – 0.19	Weak
0.20 – 0.39	Moderate
0.40 – 0.59	Relatively strong
0.60 – 0.79	Strong
0.80 – 1.00	Very strong

2.5. Atmospheric transport modeling

The European Centre for Medium-Range Weather Forecast (ECMWF) supplies global wind data employed by the International Data Centre (IDC) of the CTBTO to monitor the movement of radionuclides resulting from nuclear detonations towards the monitoring stations. The IDC employs the FLEXPART model to calculate particle trajectories, aiding in comprehending atmospheric dispersion [21]. This software creates visual representations known as "Fields-of-Regard" (FOR), shown in Figure 3. These maps illustrate the potential sources of air and radionuclides at various timestamps during sampling. These FOR maps assist in identifying the source, release strength, and path taken by radionuclides to monitoring stations upon their detection.

On the other hand, the diagram in Figure 4 is generated using HYSPLIT with 240 hours of backward trajectory. These tools help to understand wind direction during SWM season in Malaysia. The choice of the 240-hour duration aligns with the limit set by the software provider. At the same time, it appears logical that extending the time frame further might not yield substantial benefits due to the potential rise in error rates.

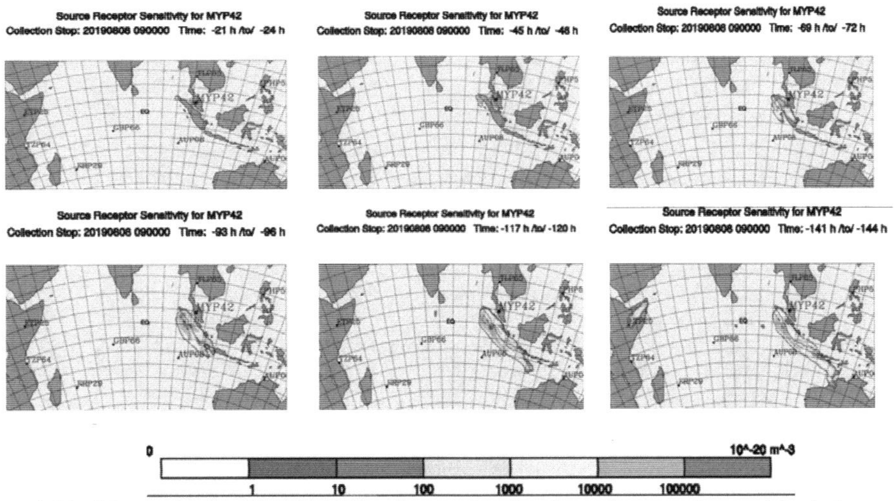

Figure 3. Field-of-Regard using FLEXPART software showing the trajectory of potential wind source.

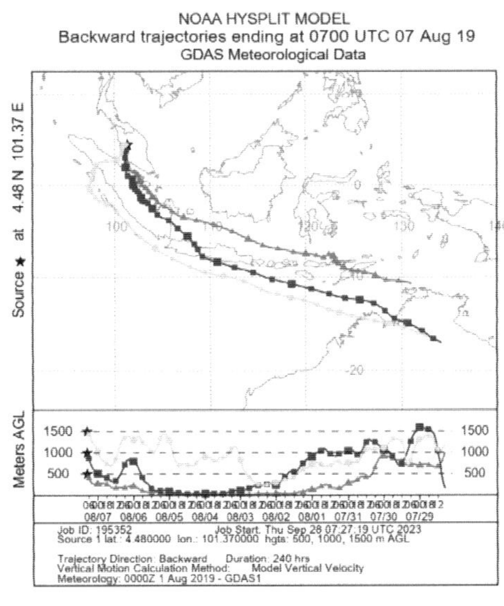

Figure 4. HYSPLIT trajectory with 240 hours backward trajectory during SWM.

3. Results and discussion

Even though longer backward trajectories might suffer from reduced accuracy owing to accumulated errors in meteorological data and the model's simplifications over time, this observation is contrasted by the prevailing southwest wind direction in the study area during the SWM, as evidenced in the Flexpart model's Figure 3 and the HYSPLIT trajectory's Figure 4. This finding corroborates the conclusions of prior studies [22] and [23].

3.1. Seasonal characteristics of the Be-7

This study focuses on the characteristics of Be-7 during both the SWM and non-SWM seasons to investigate its seasonal behavior, especially for the SWM. Understanding the Be-7 fluctuations exhibited during different seasons enhances our ability to comprehensively examine its origins, transport mechanisms, and interactions in the surrounding atmospheric and environmental contexts. Meticulous data preprocessing and structuring were conducted to reveal the distinct seasonal patterns of Be-7, with a specific emphasis on SWM. This thorough analysis facilitated the identification of fluctuation patterns that provide insights into its dynamics. Notably, the minimum, average, and maximum Be-7 values observed during both SWM and non-SWM seasons significantly contribute to our comprehensive understanding of its overall behavior. Figure 5 summarizes these Be-7 values during SWM seasons in Cameron Highlands from 2011 until 2021, while Figure 6 shows a monthly boxplot for Be-7 in the same period focusing on a Be-7 concentration during SWM only.

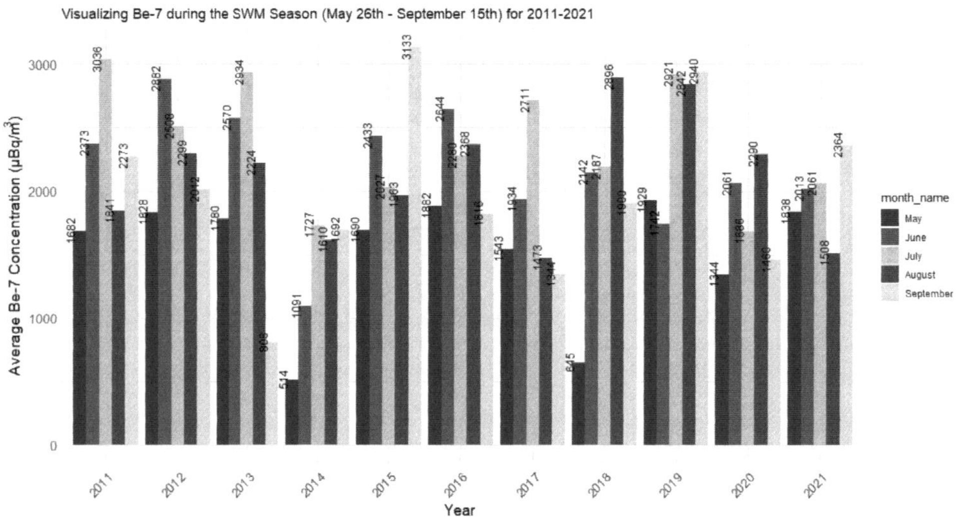

Figure 5. Summary of Be-7 during SWM for 2011 until 2021.

According to the bar graph, the Be-7 levels in May for 2014 and 2018 are lower than in other years. However, this study does not delve into the specifics behind these lower readings.

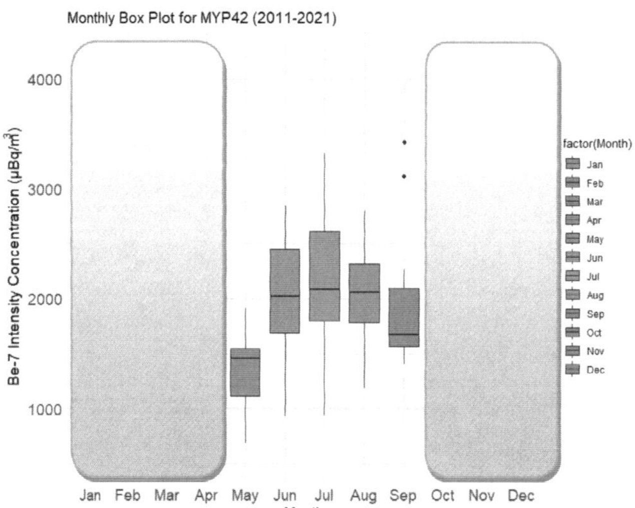

Figure 6. Monthly boxplot for MYP42 during 2011-2021, precisely during SWM months

Considering that rain or precipitation is the most influential factor in Be-7 deposition, as described in studies such as [24], [25], [26], [27], and [28], these findings do not align with the study by [3] which suggests that June experiences the least rainfall, and August sees the highest precipitation during SWM period. It may be due to the position of Cameron Highlands, which is not so affected by the SWM as mentioned in [27], where geographical attributes including latitude, longitude and altitude can give effect to radionuclide deposition rates. It could also be an indicator to a rainfall after certain period at other location. However, studies on those matters are not included in this article.

3.2. The correlation of Be-7 with other meteorological parameters

This analysis highlights the nuanced variations in correlations between Be-7 and various meteorological parameters, with some parameters showing consistent patterns and others demonstrating subtle shifts in correlation strength. The correlation result is depicted in Figures 7(a) and 7(b). MYP42_ori is the value for Be-7 and "CH" for each parameter representing Cameron Highlands.

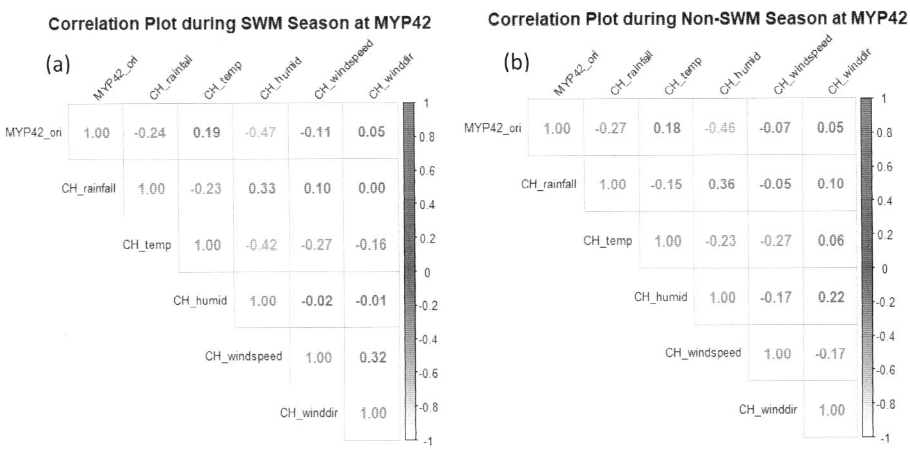

Figure 7. Correlation plot for meteorological parameters and Be-7 during SWM season (a) and non-SWM season (b).

The correlation with wind direction is consistently in the no or negligible correlation range. However, it is worth noting that the correlation with wind direction remains positive, indicating a persistent but weak positive relationship. On the other hand, wind speed, which is categorized as having no or negligible correlation during non-SWM periods, experiences a slight strengthening of negative correlation during SWM, shifting from -0.07 to -0.11. Concerning the correlation between Be-7 and rainfall, a minor reduction was observed during the southwest monsoon (SWM) season, transitioning from -0.27 to -0.24.

Despite this slight weakening, it still falls within the moderate category of negative correlation. This observation suggests a nuanced shift in the relationship between Be-7 and rainfall during SWM, but the overall strength of the negative correlation remains moderate. As for the correlation with temperature and humidity, both parameters exhibit subtle changes. Humidity during SWM shifts from -0.46 during non-SWM seasons to -0.47 during SWM, placing them in the relatively strong negative correlation category. Temperature also shows a minor difference, with a correlation of 0.19 during SWM and 0.18 during non-SWM seasons, both of which fall within the weak correlation range.

3.3. Two Sample T-Test

Due to the unequal sample sizes in SWM (n=1243) and non-SWM (n=2775), we conducted two-sample t-tests to examine the relationship between Be-7 intensity during the SWM and non-SWM periods in Tanah Rata, Cameron Highlands (MYP42).

The test's null hypothesis is rejected when the p-value falls below the significance level, set at alpha = 0.05. This suggests sufficient evidence to conclude a disparity in the mean concentration of Be-7 intensity between the two periods being compared. The obtained p-value, 2.810321e-10, is smaller than alpha, and the t-statistic value stands at 6.334705. These results indicate a significant difference in Be-7 intensity between the SWM and non-SWM periods.

4. Conclusion

In this investigation, the study examines the temporal fluctuations of the naturally occurring radionuclide Be-7 in the surface air of Tanah Rata, located in Cameron Highlands, Malaysia, specifically during the Southwest Monsoon (SWM) seasons. The study shows a negative correlation between rain and humidity during the Southwest Monsoon season, and it aligns with findings from other research on these parameters. The results show that the activity concentrations of Be-7 varied from month to month during the monsoon season between 2011 and 2021. During the Southwest Monsoon (SWM) season, the concentration of Be-7 intensity in Cameron Highlands exhibits its lowest levels in May and peaks in July. On the other hand, although the correlation coefficient values between the selected meteorological parameters exhibit relatively noticeable variations between the SWM and non-SWM seasons, they remain relatively stable for Be-7 intensity. This gives more confidence that the study location, Cameron Highlands, which is at an altitude of 1545m above sea level and part of the Titiwangsa range, has no direct effect of the southwest monsoon like the areas in the northwest and southwest regions of Peninsular Malaysia. Future studies aim to investigate what occurred in May 2014 and May 2018. Can this indicate earlier heavy rain in the affected areas during the Southwest Monsoon season?

5. References:

[1] Suhaila J, Deni S M, Zawiah Zin W A N and Jemain A A 2010 Trends in Peninsular Malaysia rainfall data during the southwest monsoon and northeast monsoon seasons: 1975-2004 *Sains Malaysiana* **39**(4). 533–542.

[2] Chenoli S N, Jayakrishnan P R, Samah A A, Hai O S, Ahmad Mazuki M Y, and Lim C H 2018 Southwest monsoon onset dates over Malaysia and associated climatological characteristics *J. Atmos. Solar-Terrestrial Phys.* **179** 81–93 o doi: 10.1016/j.jastp.2018.06.017.

[3] Vincent T and Yusoff N 2022 Trends in Peninsular Malaysia rainfall during the southwest monsoon using degree of rainfall amount (Dora) *Int. J. Infrastruct. Res. Manag.* **10**(2) 66–73 https://iukl.edu.my/rmc/publications/ijirm/.

[4] Wong C, Liew J, Yusop Z, Ismail T, Venneker R, and Uhlenbrook S 2016 Rainfall characteristics and regionalization in Peninsular Malaysia based on a high resolution gridded data set *Water* **8**(11) 500, doi: 10.3390/w8110500.

[5] Lal D and Suess H E 1968 The radioactivity of the atmosphere and hydrosphere *Annu. Rev. Nucl. Sci.* **18** (1) 407–434 doi: 10.1146/annurev.ns.18.120168.002203.

[6] Andres P 2018 Determination of atmospheric concentration of Beryllium-7 at ground level *Radiat. Prot. Environ.* **41**(3) 148 doi: 10.4103/rpe.RPE_35_18.

[7] Yoon Y Y, Koh D C, Lee K Y and Cho S Y 2016 Seasonal variation of ^{7}Be and ^{3}H in Korean ambient air and rain *J. Radioanal. Nucl. Chem.* **307**(3) 1629–1633 doi: 10.1007/s10967-015-4340-x.

[8] Ioannidou A and Papastefanou C 1994 Beryllium-7 Concentrations in the lower Atmosphere at the region of Thessaloniki (40 ° N) *Hell. Nucl. Phys. Soc. Adv. Nucl. Phys.* **5** 185–195 doi: https://doi.org/10.12681/hnps.2902.

[9] Chae J S and Kim G 2019 Large seasonal variations in fine aerosol precipitation rates revealed using cosmogenic ^{7}Be as a tracer *Sci. Total Environ.* **673** 1–6 doi: 10.1016/j.scitotenv.2019.03.482.

[10] Seif D R and Koleleni Y I 2020 Influence of meteorological parameters on the variations of atmospheric radionuclides concentration in Dar es Salaam, Tanzania *Asian J. Res. Rev. Phys.* **3**(3) 1–18 doi: 10.9734/ajr2p/2020/v3i330120.

[11] Bianchi S *et al.* 2019 Analysis of trends, periodicities, and correlations in the Beryllium-7 time series in Northern Europe *Appl. Radiat. Isot.* **148** 160–167 doi: 10.1016/j.apradiso.2019.03.038.

[12] Kusmierczyk-Michulec J, Gheddou A and Nikkinen M 2015 Influence of precipitation on ^{7}Be concentrations in air as measured by CTBTO global monitoring system *J. Environ. Radioact.* **144** 140–151 doi: 10.1016/j.jenvrad.2015.03.014.

[13] Houali A, Azahra M, El Bardouni T, Ferro García M A, Piñero García F and Chham E 2019 Impact of the

meteorological parameters on the behaviour of 7Be at ground level in Tetouan city, Morocco from June 2015 to February 2017 *J. Radioanal. Nucl. Chem.* **322**(2)271–280 doi: 10.1007/s10967-019-06723-2.

[14] Liu G, Wu J, Li Y, Su L and Ding M 2020 Temporal variations of ^{7}Be and ^{210}Pb activity concentrations in the atmosphere and aerosol deposition velocity in Shenzhen, South China *Aerosol Air Qual. Res.* **20**(7) 1607–1617 doi: 10.4209/aaqr.2019.11.0560.

[15] Zalewska T, Biernacik D, and Marosz M 2020 Correlations between ^{7}Be, ^{210}Pb, dust and PM10 concentrations in relation to meteorological conditions in northern Poland in 1998–2018 *J. Environ. Radioact.* **228** 106526 Mar. 2021 doi: 10.1016/j.jenvrad.2020.106526.

[16] Zalewska T and Biernacik D 2022 Be-7 and Pb-210 in fallout and aerosols in 2000–2016 in central Europe — Deposition velocity and dependence on meteorological parameters *Sci. Total Environ.* **826** 154205 doi: 10.1016/j.scitotenv.2022.154205.

[17] Bas M C, Ortiz J, Ballesteros L and Martorell S 2007 Analysis of the influence of solar activity and atmospheric factors on ^{7}Be air concentration by seasonal-trend decomposition *Atmos. Environ.* **145** 147–157 Nov. 2016 doi: 10.1016/j.atmosenv.2016.09.027.

[18] Zheng M. *et al.* 2021 Solar and meteorological influences on seasonal atmospheric 7Be in Europe for 1975 to 2018 *Chemosphere* **263** 128318 doi: 10.1016/j.chemosphere.2020.128318.

[19] Rashid F I A and Zolkaffly M Z A 2021 Multi-year study of natural radionuclides Beryllium-7 (Be-7) activity concentrations in surface air in Tanah Rata, Cameron Highlands, Pahang, *IOP Conf. Ser. Mater. Sci. Eng.* **1106**(1) 012029 doi: 10.1088/1757-899X/1106/1/012029.

[20] Liyew C M and Melese H A 2021 Machine learning techniques to predict daily rainfall amount *J. Big Data* **8**(1) 53 doi: 10.1186/s40537-021-00545-4.

[21] Kuśmierczyk-Michulec J and Bourgouin P 2018 Influence of mineral dust on changes of 7Be concentrations in air as measured by CTBTO global monitoring system *J. Environ. Radioact.* **192** 454–466 doi: 10.1016/j.jenvrad.2018.07.015.

[22] Satari S Z, Zubairi Y Z, Hussin A G and Hassan S F 2015 Some statistical characteristic of Malaysian wind direction recorded at maximum wind speed: 1999-2008 *Sains Malaysiana* **44**(10) 1521–1530 doi: 10.17576/jsm-2015-4410-18.

[23] . Mokhtar N. A *et al.*, "Modelling wind direction data of Langkawi Island during Southwest monsoon in 2019 to 2020 using bivariate linear functional relationship model with von Mises distribution," *J. Phys. Conf. Ser.*, vol. 1988, no. 1, 2021, doi: 10.1088/1742-6596/1988/1/012097.

[24] Ajtic J, Sarvan D, Djurdjevic V, Hernández-Ceballos M, and Brattich E 2017 Beryllium-7 surface concentration extremes in Europe *Facta Univ. - Ser. Physics, Chem. Technol.* **15**(1) 45–55, , doi: 10.2298/FUPCT1701045A.

[25] Kremenchutskii D A 2021 Influence of precipitation on the daily beryllium-7 (7Be) activity concentration in the atmospheric surface layer *J. Environ. Radioact.* **237** 106722 doi: 10.1016/j.jenvrad.2021.106722.

[26] Neroda A S,. Goncharova A A, Goryachev V A, Mishukov V F and Shlyk N V 2016 Long-range atmospheric transport Beryllium-7 to region the Sea of Japan *J. Environ. Radioact.* **160** 102–111 doi: 10.1016/j.jenvrad.2016.04.030.

[27] Morikawa K and Sugihara S 2021 Seasonal variations in atmospheric deposition of radioactive Beryllium and lead associated with atmospheric removal processes at Fukuoka, Japan *J. Junshin Gakuen Univ.* **11**.

[28] Kremenchutskii D A and Konovalov S K Beryllium-7 (^{7}Be) and its variability in the near-surface atmosphere of Crimea, the Black Sea region *Atmos. Pollut. Res.* **13**(5)101406, 2022, doi: 10.1016/j.apr.2022.101406.

Acknowledgements

The authors would like to express their sincerest gratitude to the Government of Malaysia's Department of Public Service for the financial support provided through the scholarship grant, Dr. Mohd Hisham Mohd Anip - Deputy Director General (Strategic & Technical) of the Malaysian Meteorological Department, for his inputs and consultancies and for providing substantial references on the Malaysian climate. The authors also gratefully acknowledge the NOAA Air Resources Laboratory (ARL) for providing the READY website (http://www.ready.noaa.gov) used in this study.

IOP Conf. Series: Materials Science and Engineering 1308 (2024) 012020 doi:10.1088/1757-899X/1308/1/012020

CTBT data for environmental radiological monitoring: a preliminary study in Malaysia

B Baharuddin[1,a)], M W Yii[1,b)], N F Yusof[1,c)], N A M Jaffary[1,d)], A H C Kamaruddin[1,d)], A I A Bakar[2,e)], A S Z M Halmi[2,f)], M S M. Sanusi[3,g)] and J S Johan[3,h)]

[1]Malaysian Nuclear Agency, Bangi, 43000 Kajang, Selangor

[2]International Islamic University Malaysia, Bandar Indera Mahkota, 25200 Kuantan, Pahang Darul Makmur,

[3]University Teknologi Malaysia, 81310 Skudai, Johor Bahru

[a]shilla@nm.gov.my, [b]yii@nm.gov.my,[c]ibraheem2507@gmail.com, [d]fadzilah@nm.gov.my, [e]assyikeen@nm.gov.my, [f]shakir.zufayri3399@gmail.com, [g]hadi@nm.gov.my, [h]joyce@graduate.utm.my

Abstract. Numerous anthropogenic nuclear activities in the Asia Pacific have exposed the region to radiological contamination risks, dating back to nuclear testing in 1946, with the most recent nuclear tests by North Korea. Additionally, the region had faced two major nuclear reactor disasters: the 1986 Chernobyl incident and the 2011 TEPCO Fukushima Nuclear Power accident. The latter led to the release of long-lived fission products from nuclear reactors, primarily volatile iodine and caesium isotopes into the atmosphere. These events have raised significant concerns for the international community, particularly neighbouring countries like Malaysia. Understanding atmospheric processes is essential for assessing the potential pathways of anthropogenic radionuclide transport and its impacts. Which later can be used to devise effective monitoring and mitigation strategies. This study, part of an environmental radiological monitoring initiative, utilises data from the CTBTO database. It aims to create a comprehensive mapping of anthropogenic radionuclide concentrations, focusing on Americium-241 (^{241}Am), Caesium-137 (^{137}Cs), Strontium-90 (^{90}Sr), and Plutonium isotopes ($^{239+240}$Pu) fallout in the Malaysia-South China Sea region. However, according to the CTBTO database, relevant anthropogenic radionuclide detected more than three years after the Fukushima incident is ^{137}Cs. Employing ArcGIS and CTBTO data, the study examines the influence of the Northeast and Southwest Monsoons on ^{137}Cs concentration mapping. these findings contribute to a global understanding of radionuclide distribution, with initial results showing the presence of detected anthropogenic radionuclide ^{137}Cs in the Malaysian environment. The 2014 pattern became a piece of preliminary evidence that the Northeast Monsoon had slightly elevated concentrations in the Malaysian environment, indicating its influence on radionuclide movement

1. Introduction

Many nuclear activities have put the Asia Pacific region at risk of radiological contamination since 1946 because it was used as a site for nuclear tests which last test was performed in 2017 by North Korea. The region also encountered two major reactor accidents: the 1986 Chornobyl and the 2011 Tokyo Electric Power Company's (TEPCO) Fukushima Nuclear Power accident. These major events had released fission radionuclides into the environment and pose major concern to the international community, especially to the nearby countries, including Malaysia. Due to the 2011 TEPCO Fukushima Nuclear Power accident, there were releases of long-lived medium or low-volatility fission products from the NPP reactors, but the atmospheric release was dominated by volatile isotopes of iodine and caesium (Shaat, Abdelhady, Rowayda, & Aboualo, 2019). Following this, several questions have been raised among scientists in Malaysia:

i. Are these anthropogenic radionuclides able to reach and accumulate in the Malaysian environment?
ii. To develop a map of ^{137}Cs distribution that presentable and easy to understand by the policy maker and public.

Factors such as atmospheric and hydrological processes is significant for assessing the potential pathways and impacts of anthropogenic radionuclide transport, as well as for developing effective strategies for monitoring and mitigating their environmental and health consequences. This is because once radionuclides are released into the atmosphere, natural processes such as wind patterns, precipitation, evaporation, and water currents can influence their transport over short or long distances. This study is part of the initiative for environmental radiological monitoring and understanding anthropogenic radionuclide transportation through information given by The Comprehensive Nuclear-Test-Ban Treaty Organisation (CTBTO) database. This study is also an initiative to develop a comprehensive radiological monitoring database and mapping, specifically on anthropogenic radionuclides released from man-made activities such as nuclear reactors and nuclear weapon tests to incorporate it with the National Disaster Management Plan/Program. This is due to the abovementioned disastrous events and ambiguity in the global current situation such as (i) increasing interest in nuclear power among neighbouring countries, (ii) conflict in international politic that motivates country to develop nuclear weapons and leading to the nuclear test such as by the Democratic People's Republic of Korea (DPRK), India, and Pakistan; and (iii) nuclear power plant accident beyond human control such as the 2011 TEPCO Fukushima Nuclear Power accident that caused by nature. Using ArcGIS software and CTBTO data, this study aims to create a comprehensive mapping of anthropogenic radionuclide concentrations and examine their relationship with natural processes such as wind patterns, precipitation, evaporation, and water currents. Specifically, the focus will be on mapping the fallout of Americium-241 (^{241}Am), Caesium-137 (^{137}Cs), Strontium-90 (^{90}Sr), and Plutonium isotopes ($^{239+240}$Pu) in the region encompassing Malaysia and the South China Sea. This digital mapping database will contribute to the global understanding of the distribution and concentration patterns of these radionuclides. Currently, the preliminary findings indicate the presence of ^{137}Cs as the detected anthropogenic radionuclide reaching Malaysia. The concentration mapping will be developed considering the influence of the Northwest Monsoon and Southwest Monsoon seasons.

This study aims to investigate and discover scientific evidence as to whether fallout from the previous Asia Pacific nuclear incidents/disasters has reached Malaysia's environment by creating a anthropogenic radiological mapping, particularly for radionuclides released from man-made activities such as nuclear reactors and nuclear weapon tests, integrating it with National Disaster Management Plans/Programs. This study also attempts to establish modelling that provides the ability to assess the current and future trends of radionuclide activities in the Malaysian atmosphere in order to enable relevant policy decisions to be made. Therefore, atmospheric radionuclide activity models play an important role in providing information for better and more efficient air quality management planning. An effective air quality management system must be able to provide the authorities with information about the current and likely future trends, throughout the area enabling them to make necessary assessments regarding the extent and type of the air pollution control management strategies to be followed [7].

1.1. Democratic People's Republic of Korea (DPRK) nuclear tests

In the 21st century, the DPRK is the only country to have been recorded doing nuclear tests. Since its first test in 2006, North Korea has done a total amount of five other tests, which were conducted in the year 2009, 2013, twice in 2016, and 2017. The reasons for these tests are mostly political, as it is known that there have been multiple talks between the country and other powerful nations such as the United States, China, and Japan, with no definite conclusion being achieved. The past couple of decades also have seen at least two major crises and a continuous deterioration of U.S.-North Korean relations (Liu Lin, 2006). However, the consequences of such nuclear tests are far from being solely political. The nuclear tests raise an alarm about the environmental repercussions and ways to handle them, given that it has been known that nuclear explosions never had a good track record environmentally. The tests were confirmed with the detection of seismic activity, infrasound observation, remote-sensing studies, and measurement of anthropogenic radionuclides and noble gases by various CTBTO detection stations, national institutes, and meteorological agencies that are close enough to the location (Gaebler et al., 2019).

1.2. Fukushima TEPCO's Nuclear Accident
On March 11, 2011, Japan was hit by the Tohoku earthquake, which resulted in a massive tsunami that caused an unprecedented release of anthropogenic radionuclides from the Fukushima Daiichi nuclear power plant to the Northwest region of the Pacific Ocean (Buesseler et. al., 2012). Reactor failures at the nuclear power plant led to the direct release of radionuclides into the sea and the deposition of atmospheric releases, with the direction of the prevailing winds at the time of the accident was towards the sea (J. Vives i Batlle et. al., 2014). Initial investigations by the International Atomic Energy Agency (IAEA) also confirmed that the radionuclides from the accident were released from the power plant to the atmosphere and deposited on land and ocean. Therefore, the fission product release mechanism and the atmospheric, marine, and terrestrial dispersion pathways need to be studied to know the extent of the radioactive fallout over Malaysia several years after the accident.

1.3. Study objectives
The importance of this study is due to the persistence of anthropogenic radionuclides in nature as their half-life can be up to several years, which may directly and indirectly have adverse consequences on life in general. The spread pathway can be determined and interpolated over the Asia Pacific region, including Malaysia by assessing the radionuclide distribution. While the anthropogenic radiological contamination affects the air, land, and sea, the focus of this project will be on the assessment of the ^{137}Cs distribution and transportation in the atmosphere following the recent nuclear incidents, which are the 2011 TEPCO Fukushima nuclear accident and the DPRK's nuclear weapon tests. This study will provide scientific evidence on the spread of anthropogenic radionuclides, as well as act as a preliminary investigation on radionuclide atmospheric transport before a comprehensive radionuclide database and movement simulation can be developed. Therefore, this study has three distinct objectives:

i. To explore CTBTO IMS data use for National Radiological Mapping and Database on common fission products to support National Disaster Management.
ii. To investigate and discover scientific evidence whether fallout from the previous Asia Pacific nuclear incidents/disasters has reached/spread out into Malaysia's environment.

2. Methodology

2.1. Collection of airborne anthropogenic radionuclides
Airborne radionuclides exist in two forms: particulates (large enough for filtration) and gases or vapours (impossible to filter). Analysing samples is essential for assessing the presence of radionuclides suspended in the air, as exposure to and inhalation of these particles can be harmful. Radionuclides in the air can stem from natural sources, such as volcanic eruptions (natural radionuclides), or human-made events like nuclear tests and nuclear power plant accidents (anthropogenic radionuclides). The concentration of airborne radionuclides can be determined for a specific radionuclide or as gross alpha or beta concentration. Worldwide, CTBTO radionuclide monitoring stations play a pivotal role in collecting data on airborne radioactive materials. These stations are equipped to detect both radionuclide particulates and gases, particularly noble gases, and then transmit this data to the International Data Centre (IDC). Key air sampling equipment includes an air filter for particulate collection, a flow meter to measure sampling flow rate, an elapsed timer to track sampling duration, and a pump to circulate air through the sampler.

IOP Conf. Series: Materials Science and Engineering 1308 (2024) 012020 doi:10.1088/1757-899X/1308/1/012020

For this study, the CTBTO IMS Radionuclide Data [radioxenon, ^{241}Am, ^{137}Cs, ^{90}Sr, ^{239}Pu, and ^{240}Pu] will be obtained from several IMS Radionuclide Station, namely as below:

RN38 – Takasaki, Japan
RN37 – Okinawa, Japan
RN52 – Tanay, Philippines
RN42 – Cameron Highlands, Malaysia
RN65 – Bangkok Thailand

In this study, the CTBT IMS radionuclide data is used to observe if these radioactive particulates are moving toward Malaysia. Data collected was before and after the Fukushima nuclear accident. Interest anthropogenic radionuclide for the project is ^{137}Cs, radioxenon, ^{241}Am, ^{90}Sr, ^{239}Pu, and ^{240}Pu. Later these data were analysed to study the pattern of ^{137}Cs concentration against meteorology data such as wind direction. ^{137}Cs was studied because of its half-life longer than other anthropogenic radionuclide sourced from reactor accident and nuclear tests. For example, half-life Iodine-131 is only 8 days. Also, the accident had resulted in 2.7 PBq of radio caesium (^{137}Cs) which contaminated forests, agricultural lands, grasslands, and urban areas, which subsequently migrated through soil and waterways in the Fukushima Prefecture [5]. This study will investigate whether ^{137}Cs subsequently migrated to Malaysia's environment through the atmosphere under the influence of the monsoon. This is after we learned that ^{137}Cs bonds with chlorides to make a crystalline powder, it can react in the environment like table salt (sodium chloride), so caesium moves easily through the air, it can dissolve easily in water and caesium binds strongly to soil and concrete, but does not travel very far below the surface. Plants and vegetation growing in or nearby contaminated soil may take up small amounts of ^{137}Cs from the soil. Small quantities of ^{137}Cs can be found in the environment from nuclear weapons and from nuclear reactor accidents [12].

The mapping and the autocorrelation analysis were conducted with ArcGIS software 9 [version 10.2 ESRI Inc. Redlands]. The data was divided into two seasons: the northeast monsoon and the southwest monsoon, to see how these monsoons influence the distribution of the particulates. The transport of radionuclides and other substances in the Earth's atmosphere is governed by the large-scale flow fields referred to by meteorologists as the large-scale circulation, or general circulation. Data collection took place both before and after the TEPCO Fukushima Nuclear Accident, as outlined below:

Table 1. Onset date for Northeast Monsoon.

Year represents	Start date (onset date)	End date (withdrawal date)
2009	2009/2010	2/3/2010
2010	29/11/2010	27/3/2011
2011	22/10/2011	23/3/2012
2012	18/10/2012	23/3/2013
2013	1/11/2013	3/4/2014
2014	15/11/2014	20/3/2015
2015	29/10/2015	22/3/2016
2016	14/11/2016	16/3/2017
2017	27/11/2017	4/4/2018
2018	11/10/2018	3/3/2019
2019	14/11/2019	8/3/2020
2020	11/11/2020	28/3/2021
2021	3/11/2021	13/3/2022

Table 2. Onset date for Southwest Monsoon.

Year represents	Start date (onset date)	End date (withdrawal date)
2010	23/5/2010	7/9/2010
2011	9/5/2011	5/10/2011
2012	18/5/2012	6/10/2012
2013	20/5/2013	5/9/2013
2014	3/6/2014	4/10/2014
2015	18/5/2015	13/10/2015
2016	24/5/2016	19/9/2016
2017	17/5/2017	20/9/2017
2018	6/5/2018	20/9/2018
2019	2/5/2019	24/9/2019
2020	18/5/2020	8/10/2020

The collected data is categorised as spatial data, focusing on anthropogenic radionuclide's types and their activity concentrations at various locations. The analysis is facilitated by Geographic Information System (GIS) software for mapping and in-depth examination.

2.2. Running data using GIS software

In this project, the GIS software employed is ArcGIS, with ARC Map utilised for subsequent mapping procedures. To accommodate the small data values provided by the CTBT in micro Becquerel (μBq), the concentration values are converted into becquerels (Bq). The anthropogenic radionuclide data, along with latitude and longitude details, is imported into the software using an Excel sheet. To create a point file that represents the sampling stations as point features, the file is right-clicked, designating the X-axis for longitudes and the Y-axis for latitudes.

3. Results and discussion

Due to the confidentiality of the CTBT data, the observed data cannot be written in this paper.

Table 3. Station locations for observed data.

Station code	location
JPP 37	Okinawa, Japan
JPP 38	Takasaki, Gunma, Japan
MYP 42	Tanah Rata, Malaysia
MNP 45	Ulaanbaatar, Mongolia
PHP 52	Tanay, Philippines
RUP 58	Ussuriysk, Russian Federation
RUP 60	Petropavlovsk-Kamchatskiy, Russian Federation

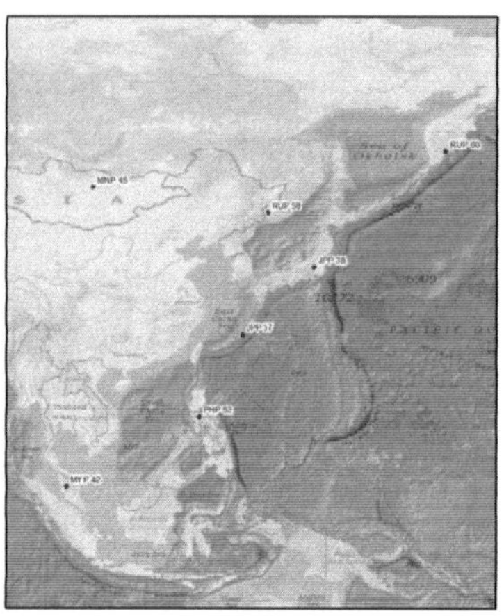

Figure 1. Station locations for observed data.

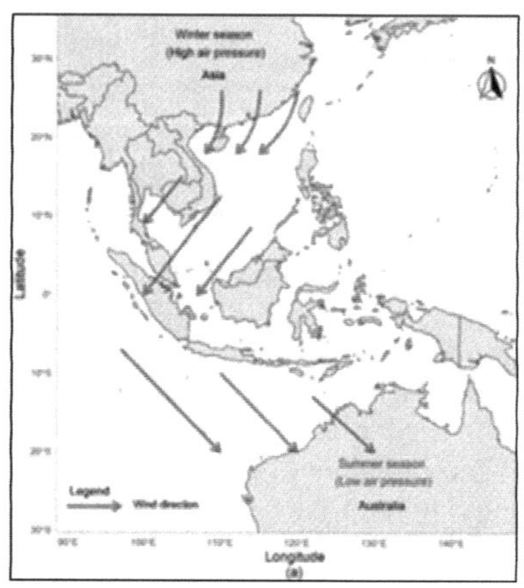

Figure 2. Wind direction during Northeast monsoon.

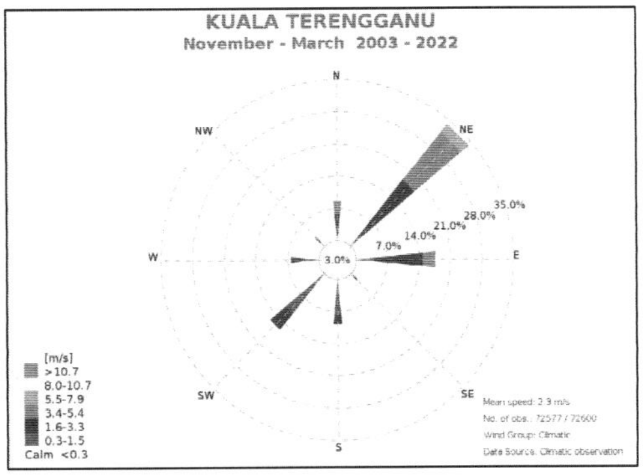

Figure 3. The windrose chart during the Northeast monsoon observed from Kuala Terengganu station demonstrates the wind direction and the wind speed during this monsoon.

Source: Malaysian Meteorological Department, 2023.

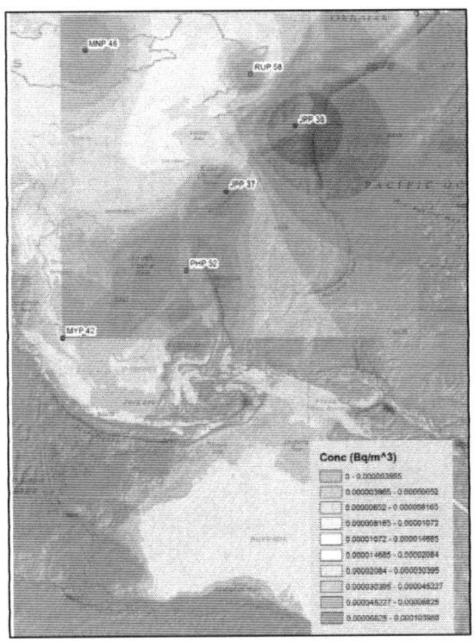

Figure 4. ^{137}Cs concentration in 2011 after the TEPCO's Fukushima Daiichi accident data observed during Northeast monsoon.

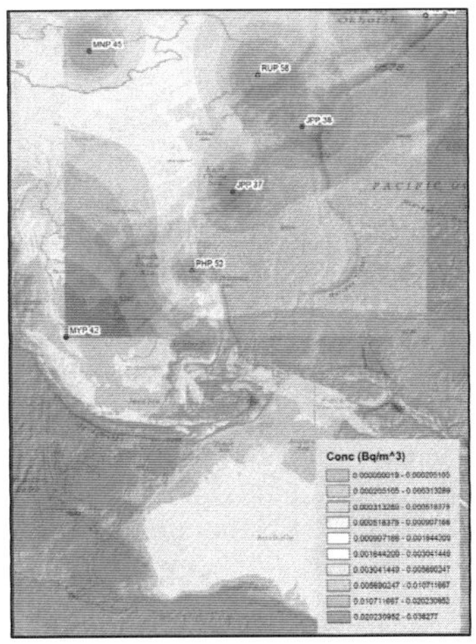

Figure 5. ^{137}Cs concentration in 2014 after the TEPCO's Fukushima Daiichi accident and 2013 DPRK's Nuclear Test data observed during Northeast monsoon.

In March 2011, through an observation by Japanese researchers, ^{137}Cs was detected with the same activity as that of ^{137}Cs [11]. Thus, for this study, only ^{137}Cs was observed. As ^{134}Cs had not been detected before the accident except during the emission period resulting from the Chernobyl accident in 1986 [2]. In 2011, there were minimal changes in the ^{137}Cs levels across different locations. The IMS Station JPP38 (Takasaki, Japan) experienced a significant increase, while the IMS Station MNP45 (Ulaanbaatar, Mongolia) showed a significant decrease. Fortunately, the Cs137 levels remained very low at the IMS Station MYP42 (Tanah Rata, Malaysia), indicating that it can be considered as background data. Later, in 2014 ^{137}Cs levels increased in the direction of Malaysia, while the IMS Stations in Japan, Mongolia, and the Philippines recorded significant decreases. For ^{137}Cs mapping for

IOP Conf. Series: Materials Science and Engineering 1308 (2024) 012020 doi:10.1088/1757-899X/1308/1/012020

year 2011, 2012, and 2013 also plotted to identify the pathway of ^{137}CS dispersion in the atmosphere. The ^{137}Cs levels at IMS Station MYP42 (Tanah Rata, Malaysia) showed a slight elevation, indicating a slightly higher concentration compared to other monitored locations. This result aligns with the wind direction depicted in Figure 2, as expected. concentration is within the defined limits, but it is observed that it is increasing yearly.

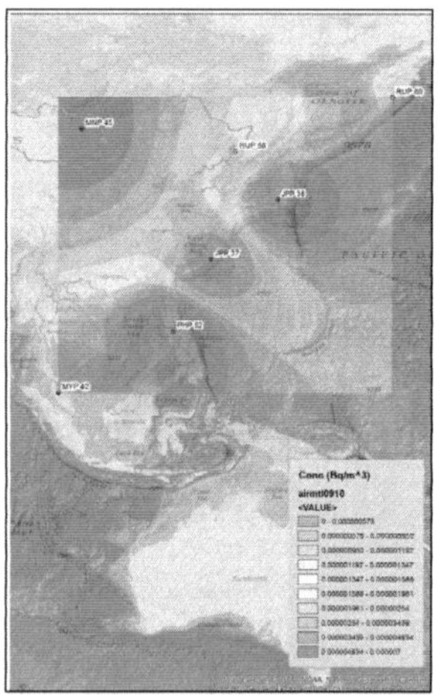

Figure 6. GIS contour map that constructed in the year 2009 to identify the pathway of ^{137}Cs dispersion in the atmosphere.

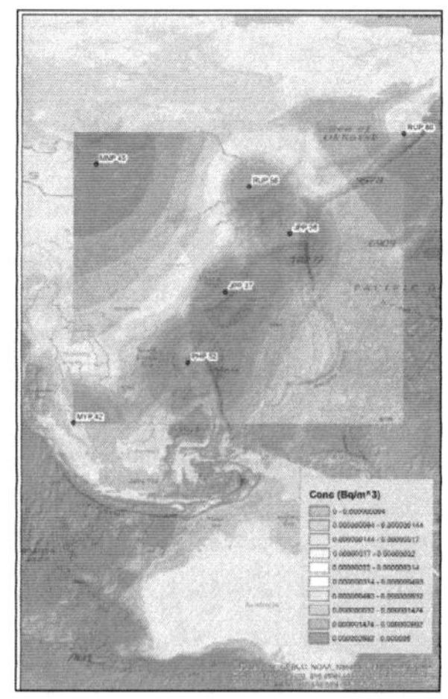

Figure 7. GIS contour map that constructed in the year 2010 to identify the pathway of ^{137}Cs dispersion in the atmosphere.

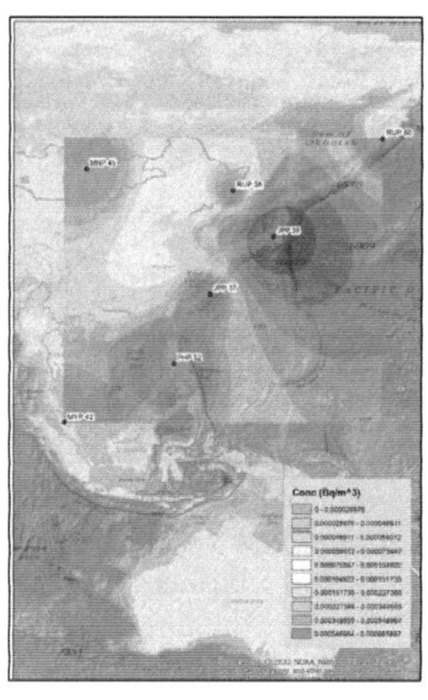

Figure 8. GIS contour map that constructed in the year 2011 to identify the pathway of ^{137}Cs dispersion in the atmosphere.

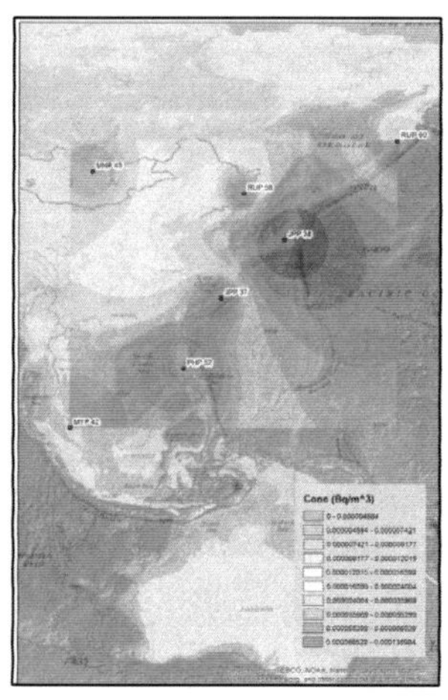

Figure 9. GIS contour map that constructed in the year 2012 to identify the pathway of ^{137}Cs dispersion in the atmosphere.

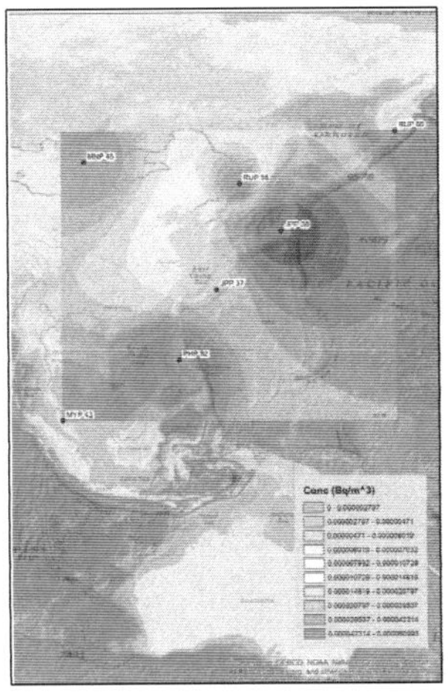

Figure 10. GIS contour map that constructed in the year 2014 on Southwest monsoon to identify the pathway of ^{137}Cs dispersion in the atmosphere.

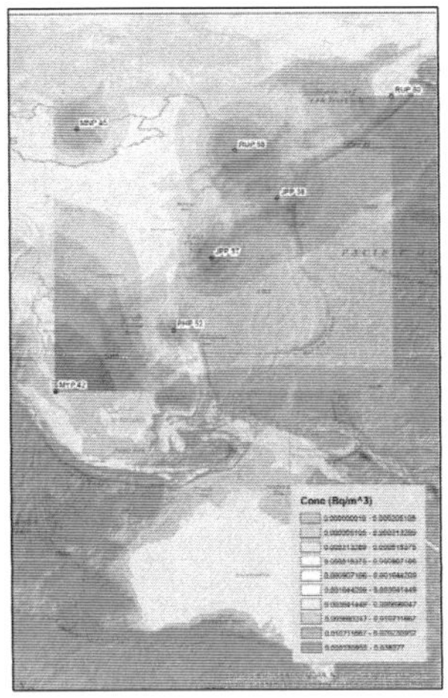

Figure 11. GIS contour map that constructed in the year 2014 on Northeast monsoon to identify the pathway of ^{137}Cs dispersion in the atmosphere.

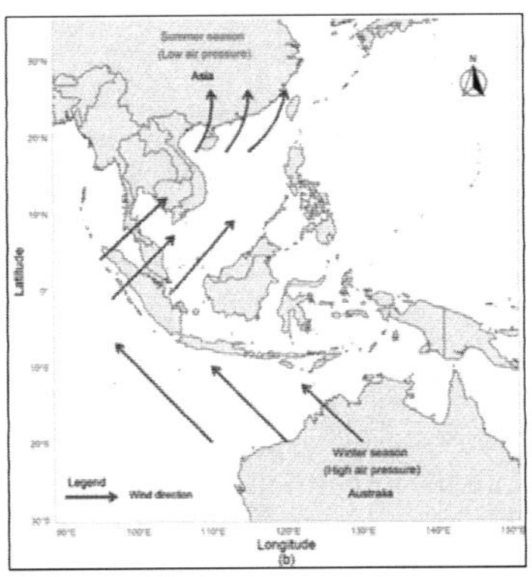

Figure 12. Wind direction during the Southwest monsoon.

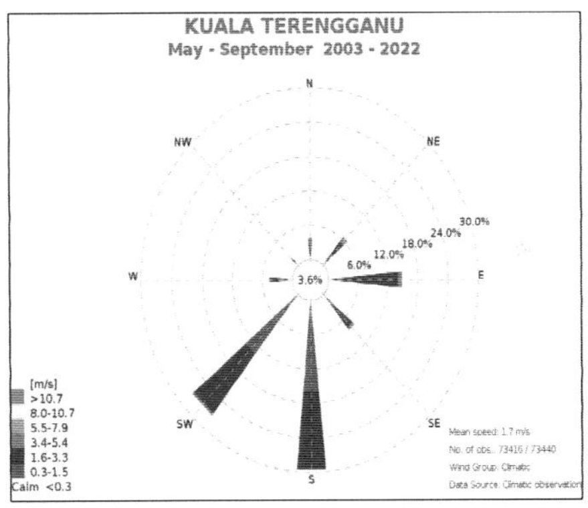

Figure 13. The windrose chart during the Southwest monsoon observed from Kuala Terengganu station demonstrates the wind direction and the wind speed during this monsoon.

Source: Malaysian Meteorological Department, 2023.

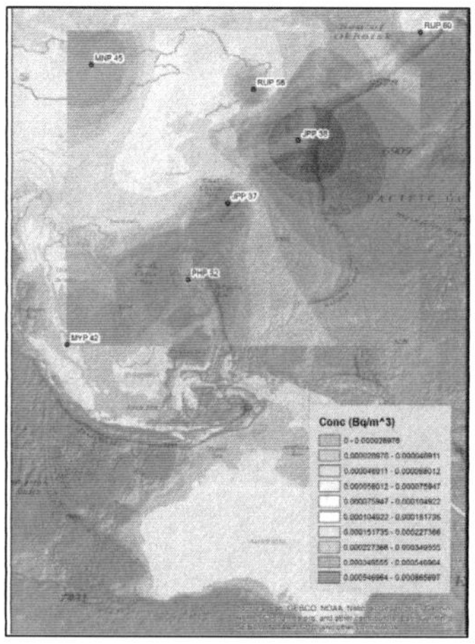

Figure 14. ^{137}Cs concentration in 2011 after the TEPCO's Fukushima Daiichi accident data observed during Southwest monsoon.

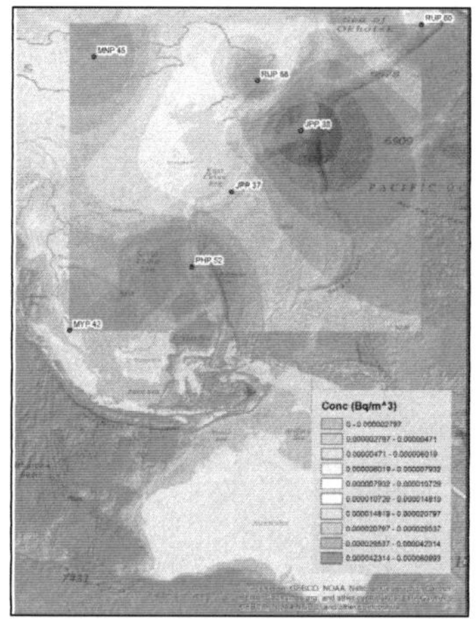

Figure 15. ^{137}Cs concentration in 2014 after the TEPCO's Fukushima Daiichi accident and 2013 DPRK's Nuclear Test data observed during Southwest monsoon.

The contour maps presented in this study illustrate that during the southwest monsoon, there were no significant changes in ^{137}Cs concentrations between year 2011-2014. However, traces of ^{137}Cs were still detected at the IMS Station MYP42 (Tanah Rata, Malaysia), albeit at extremely low levels. This situation is attributed to the prevailing wind direction during the southwest monsoon period.

4. Summary and conclusions

Despite the absence of significant nuclear reactors in Malaysia and its geographical distance from nuclear weapon states such as the USA, Russia, and China, complacency regarding radiological hazards should not be assumed. Findings from this preliminary study indicate that even with considerable distances between Malaysia and countries like Japan, DPRK, and Russia, anthropogenic radionuclides have the potential to reach Malaysia within several years following incidents or nuclear tests in these regions or neighbouring areas within the Asia-Pacific. This transport is facilitated by wind direction patterns and the prolonged half-life of these radionuclides. It is important to note that multiple incidents can contribute to the accumulation of anthropogenic radionuclides over time. A study by Japanese scientists through chemical analysis suggested that dust particles were the major carriers of ^{137}Cs during the resuspension period [2]. When chemical analysis suggests that dust particles were the major carriers of ^{137}Cs during the resuspension period, it is likely that ^{137}Cs can be easily distributed by wind direction. Dust particles are typically very light and can be carried over long distances by the wind. So, the wind direction would play a crucial role in determining the dispersion pattern of ^{137}Cs carried by these dust particles. The specific wind direction would dictate where the contaminated dust particles are transported, potentially affecting the areas where ^{137}Cs is deposited. A study at Barakah Power Plant UAE, indicated that that in hypothetical severe accident conditions, the ^{137}Cs dispersion followed the main wind direction [3]. This finding aligns with this study which the North-East monsoon has distributed the ^{137}Cs release from Japan to the Malaysian environment as detected by IMS radionuclide station in Cameron highlands, Pahang. Thus, a frequent radiological monitoring is suggested to be conducted to ensure that Malaysia is ready for any consequences.

This study proves the statement by Martin in 2019 and aligns with an observation by Thao Linh Tran in 2022 about turbulence in Southeast Asia. Martin claimed that the distribution of a radionuclide in the Earth's atmosphere is influenced by five key factors. However further study is needed to see if precipitation has a significant effect to the distribution of ^{137}Cs First the radionuclide's source, pertains to how it enters the atmosphere and where this introduction occurs. Second, by examining the extent to which the radionuclide undergoes chemical reactions or binds with airborne particles. Third, horizontal and vertical mixing processes, primarily driven by turbulence, lead to the dispersion and dilution of the radionuclide in regions distant from its origin. Fourth, both wet and dry deposition on the Earth's surface can physically remove it from the atmosphere. Lastly, radioactive decay, as characterized by the radionuclide's half-life, serves as another removal mechanism [4]. The dispersion of ^{137}Cs into the atmosphere over Malaysia is shaped by both the speed and direction of the wind, which, in turn, affect its concentration due to the mixing processes which may be driven by turbulence, this is also strongly supported by the fact that Southeast Asia is located in the western North Pacific (WNP), which is the most prolific tropical cyclone (TC) basin that producing 30% of the global total TCs accompanied by extreme winds, torrential rains, and storm surges. This led to the dispersion and dilution of the radionuclide in regions distant from its origin [8].

Finally, GIS helps in mapping the ^{137}Cs, distribution that is presentable and easy to understand by the general public and policymakers. CTBT data is good because it has a very sensitive technology which even in a very small amount of radioactivity in μBq, is detectable, and this is considered as precise data. This research constitutes an initial exploration in the field of environmental monitoring. It aims to identify the specific areas of further investigation necessary to validate the underlying assumption through a more exhaustive and in-depth study.

5. References

[1] Buesseler K O, Jayne S R, Fisher, N S, Rypina I I, Baumann H, Baumann Z, Breier C F, Douglass E M, George J, Macdonald A M, Miyamoto H, Nishikawa J, Pike S M, Yoshida S 2012 Fukushima-derived radionuclides in the ocean and biota off Japan *Proc. of the Nat. Acad. of Sci. of the United States of America* **109**(16) 5984-5988. https://doi.org/10.1073/pnas.1120794109.

[2] Hussin A A, Moslim N H, Mokhtar N A and Zubairi Y Z 2021 understanding the behaviour of wind direction in Malaysia during monsoon seasons using replicated functional relationship in von mises distribution **50**(7) 2035-2045 (Sains Malaysiana).

[3] Kinase T, Adachi K, Sekiyama T T, Kajino M, Zaizen Y and Igarashi Y 2020 Temporal variations of ^{90}Sr and ^{137}Cs in atmospheric depositions after the Fukushima Daiichi nuclear power plant accident with long-term observations *Scientific Reports* https://www.nature.com/articles/s41598- 020-78312-3.

[4] Lee J K, Kim J C, Lee K J, Belorid M, Beeley P A and Yun J L 2014 Assessment of wind characteristics and atmospheric dispersion modelling of ^{137}Cs on the Barakah NPP area in the UAE *Nucl. Eng. and Technol.* **46**(4) 557-566 https://www.sciencedirect.com/science/article/pii/S1738573315301248 .

[5] Liu, L. (2006). The North Korean Nuclear Test and Its Implications. Central Asia-Caucasus Institute Silk Road Studies Program. Uppsala. ISBN: 91-85473-28-6. Utrikes Departementet, the Swedish Foreign Ministry.

[6] Martin P and McBride J L 2019 Radionuclide behaviour and transport in the tropical atmospheric environment. in radioactivity in the environment **18** 59-91 (Elsevier) https://doi.org/10.1016/B978-0-08-045016-2.00002-3

[7] Onda Y, Taniguchi K, Yoshimura K et al. 2020 Radionuclides from the Fukushima Daiichi nuclear power plant in terrestrial systems *Nat. Rev. Earth Environ. 1* 644–660 https://doi.org/10.1038/s43017-020-0099-x.

[8] Shaat M, Abdelhady A, Rowayda A M, and Aboualo R F M 2019 Radiological impact due to atmospheric releases of the source term for F-D, Unit 1, Using HOTSPOT code *Ann. of Ecol. and Environ. Sci.* **3**(1), 33-39 https://doi.org/ 10.22259/2637-5338.0301005.

[9] Sharma N, Bhandari K, Rao P and Shukla A 2019 GIS applications in air pollution modelling https://www.geospatialworld.net/article/gis-applications-in-air-pollution-modeling/.

[10] Tran T L, Ritchie E A, and Perkins-Kirkpatrick S E 2020 A 50-year tropical cyclone exposure climatology in Southeast Asia *J. of Geophysic. Res.: Atmosph.* **127** e2021JD036301 https://doi.org/10.1029/2021JD036301.

[11] Thiessen K, Thorne M, Maul P, Pröhl G, and Wheater H 1999 Modelling radionuclide distribution and transport in the environment *Environ.l Pollut.* **100**(1-3) 151–177 https://doi:10.1016/s0269-7491(99)00090-1.

[12] Vaddiraju S 2019 Mapping aof Air Pollution Using GIS: A Case Study of Hyderabad *Int. J. of Res. in Advent Technol.*7(5)) e-ISSN 2321-9637.

[13] Vives i Batlle, J., et al. (2016). Impact of the Fukushima Accident on Marine Biota, Five Years Later. Integrated Environmental Assessment and Management, 12(4), 654–658.

[14] Wan Mahmood Z U, Yii M W, Khalid M A et al. 2018 Marine radioactivity of Cs-134 and Cs-137 in the Malaysian Economic Exclusive Zone after the Fukushima accident *(J. Radioanal. Nucl. Chem.* **318** 2165–2172 https://doi.org/10.1007/s10967-018-6306-2.

[15] United States Environmental Protection Agency 2023 *Radionuclide Basics: Cesium-137* https://www.epa.gov/radiation/radionuclide-basics-cesium-137

Acknowledgment

I want to convey my heartfelt thanks and appreciation to the Malaysia Nuclear Agency, the Malaysian Meteorological Department, and the CTBTO Preparatory Commission for their invaluable support and contributions to this project.

IOP Conf. Series: Materials Science and Engineering 1308 (2024) 012021 doi:10.1088/1757-899X/1308/1/012021

Underground utility inspection using ground penetrating radar

Tengku Sarah Tengku Amran[1,a)], Mohamad Ridzuan Ahmad[1], Amer Hazreq Haron[1], Noor Azreen Masenwat[1], Ismail Mustapha[1], Mohd Fajri Osman[1], Noryana Abd Razak[1], Mior Ahmad Khusaini Adnan[1], Akmal Haziq Rahman[2], Amir Afiq Adnan[2] and Hasimah Ali[2]

[1]Material Structure Integrity Group (NDT-MSI), Industrial Technology Division, Malaysian Nuclear Agency, 43000, Kajang, Selangor, Malaysia.
[2]Electrical Engineering Technology Faculty, Kampus Alam Pauh Putra, Universiti Malaysia Perlis (UniMAP), 02600, Arau, Perlis, Malaysia.

[a]sarah@nm.gov.my

Abstract. Underground utility refers to any below-ground line, installation, or structure used by a service or utility provider. Underground utility inspection is a process of examining the designated area that needs to be inspected in detail. The purpose of the inspection is to ensure that the underground utility is in good condition and to check for any faults. By using the ground penetrating radar (GPR), the technology can capture images below the surface of the ground using radio waves. The various data through previous references are used to discuss and analyse the capability of GPR for underground utility inspection to ensure its gives the best performance for the inspection. GPR technology has become an essential tool for locating buried utilities and preventing damage and accidents during construction projects. This paper explores various advancements in GPR technology, including data processing methods and scanning techniques, to improve the accuracy and effectiveness of underground utility mapping. The research presented in this paper highlights the abilities of GPR to inspect underground utilities. The GPR data used in this project was collected from the evaluation and verification facility (EVF) in the Malaysia Nuclear Agency. The accuracy of these predicted positions was validated from GPR data and compared with as-built drawings from the contractor. The alternative analysis approaches that could be applied to improve the GPR accuracy when the optimum conditions are not met are also discussed.

1. Introduction

Urban planning, building, and public safety all heavily depend on the underground utility inspection method. Its significance cannot be emphasized since it helps prevent accidents, reduce construction delays, and decrease service disturbance. Safety is of the highest priority, both for construction employees and for the general public. By spotting potential hazards like gas leaks, live electrical lines, and unstable ground conditions, underground utility inspection helps lower the chance of mishaps, injuries, and fatalities [1]. Accurate information of underground utilities is essential for excavation, and the proposed method using ground penetrating radar (GPR) sets a new benchmark for precise location with high accuracy [1]. Underground utility inspection helps detect and map existing beneath the surface utilities, such as water pipes, gas lines, sewage systems, and communications cables, before any construction or excavation work is started. This helps avoid unforeseen damage to these important infrastructure parts, which may be expensive to fix and cause disruptions to crucial services. Accurately locating subsurface lines before construction can help avoid expensive delays and unexpected expenses. Through careful inspection, accidental damage to utility wires can be prevented, preventing project delays, emergency repairs, and fines.

It is necessary to define GPR before digging into its history. To identify structures and changes in material characteristics within lossy dielectric materials, GPR employs electromagnetic fields to probe the materials [2]. Measurements of reflection and transmission are used, as shown in Figure 1. Up until now, natural geologic materials have seen most applications, but man-made composites like concrete, asphalt, and other building materials are also widely used where electromagnetic fields can exist in such lossy dielectric materials. A certain depth will be reached by fields before they are absorbed. The electromagnetic fields that are used in GPR spread as essentially nondispersive waves [3]. Events comparable to this occur as the signal is transmitted through the material, is dispersed, or is reflected by changes in impedance. to the signal being sent. To put it another way, signal detection is easy because a GPR antenna has two parts, a transmitter, and a receiver [4]. The transmitter antenna will generate an electromagnetic pulse that transmits below the ground. While the receiver will pick up the bock scatter and reflected waves that encounter a discontinuity in the dielectric properties (e.g. pipe, cable, sinkhole) [5]. The remaining electromagnetic waves keep going and continue to deepen the subsurface levels. The proposed antenna can be used for the detection of artifacts in the deeper ground known as ground-penetrating radar applications [6].

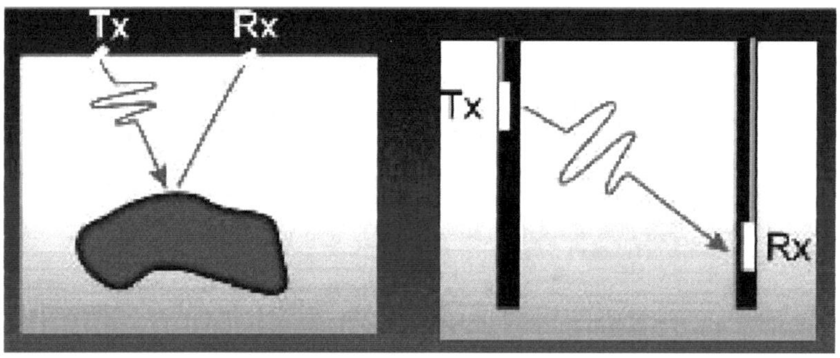

Figure 1. Ground penetrating radar uses radio waves to probe the subsurface of lossy dielectric materials. Two modes of measurement are common. In the first, detection of reflected or scattered energy is used. In the second signal, variation after transmission through the material is used to probe a structure [1].

The objective of this study is to examine the potential of GPR to inspect underground utilities. The GPR data utilized in this investigation was obtained from the Evaluation and Verification Facility (EVF) of the Malaysia Nuclear Agency. The accuracy of the predicted positions was verified using GPR data and compared with as-built drawings provided by the contractor. The experiment aimed to detect a mild steel pipe buried at a specific depth at the EVF facility. Although various underground facilities, such as fiber optic cables, electrical lines, water pipes, and culverts, are buried at this site, this experiment focused solely on detecting mild steel pipes.

2. Experimental work

This experiment was held at the evaluation and verification facility (EVF) in Malaysia Nuclear Agency. The scanning area was divided into two parts which consisted of 1st phase and 2nd phase (Figure 2). The dimensions of the scanning area were 13 m × 6 m with 22 lines profile for vertical scan and 9 lines profile for horizontal scan for the 1st phase scanning area. For the second phase scanning area was 5.9 m × 14 m with 9 profiles for the vertical scan and 18 profiles for the horizontal scan (Figure 3 -5). The GPR used was a GSSI 400MHz shielded antenna. Data was collected using

IOP Conf. Series: Materials Science and Engineering 1308 (2024) 012021 doi:10.1088/1757-899X/1308/1/012021

SIR 4000 and then processed using RADAN 7 software (Figure 6). For comparison, the contractor's as-built drawing has been used for validation purposes. The depth of mild steel pipe has been used as a reference depth (1.15 m).

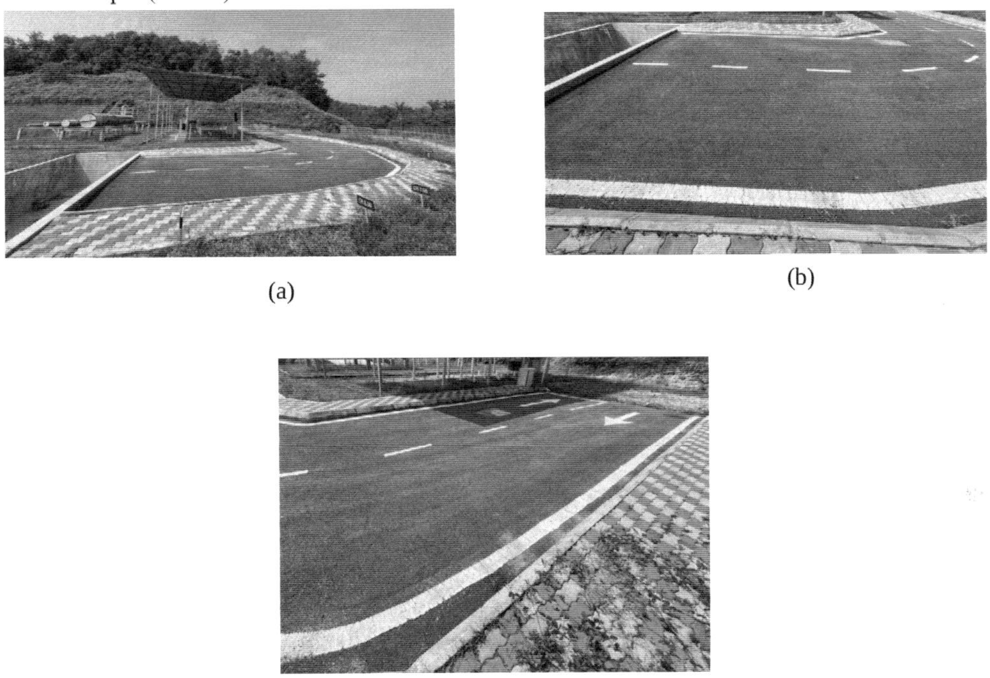

(a) (b)

(c)

Figure 2. (a) Evaluation Verification Facility (EVF) (b) First phase scanning area (c) Second phase scanning area.

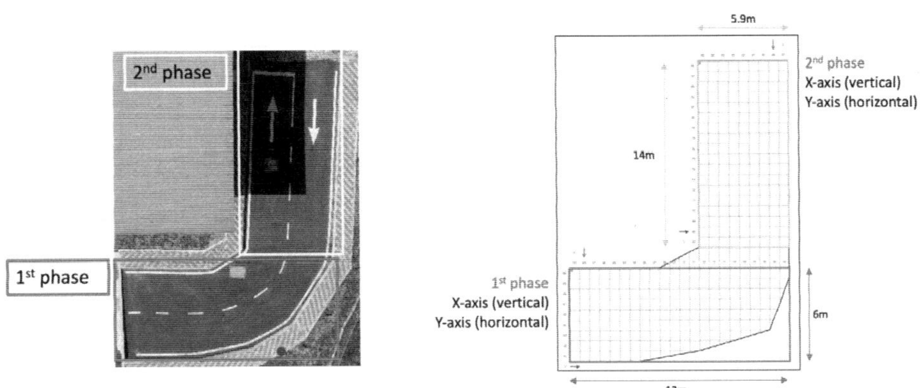

(a) (b)

Figure 3. (a) Drone images Evaluation and Verification Facility (EVF) (b) Schematic diagram for scanning area.

IOP Conf. Series: Materials Science and Engineering 1308 (2024) 012021 doi:10.1088/1757-899X/1308/1/012021

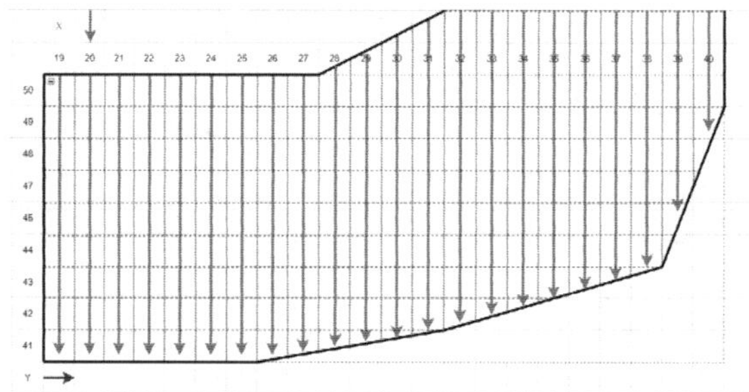

(a)

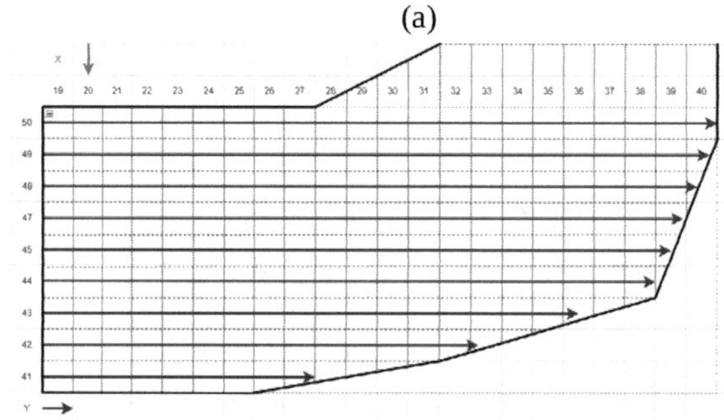

(b)

Figure 4. (a) Vertical scan 1st phase (b) Horizontal scan 1st phase

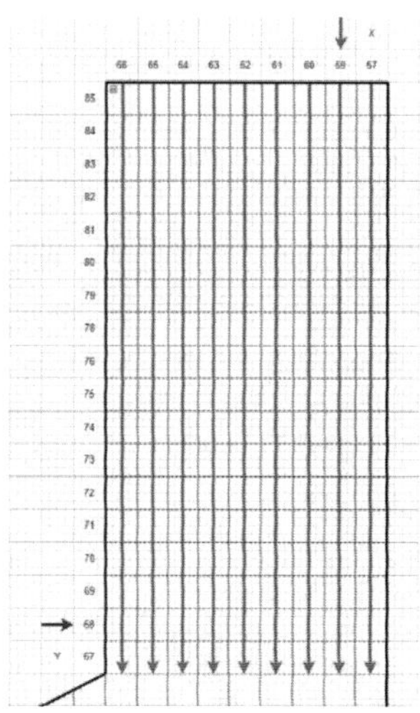

(a)

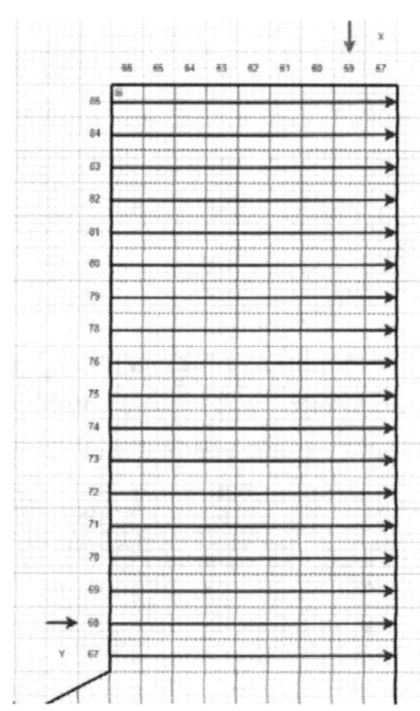

(b)

Figure 5. (a) Vertical scan 2nd phase (b) Horizontal scan 2nd phase.

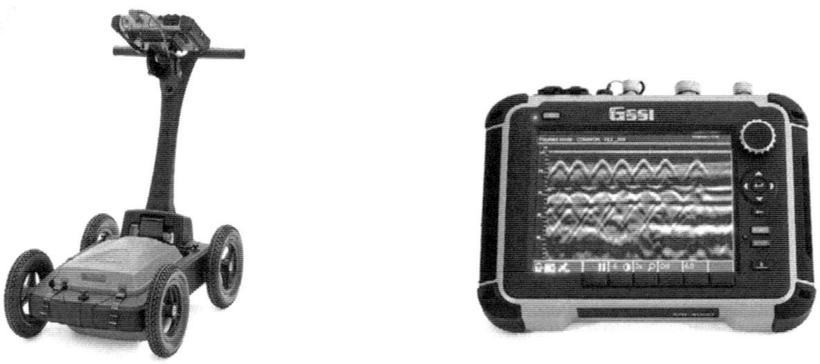

(a) (b)

Figure 6. (a) The GPR GSSI 900 MHz equipment on the GPR cart (b) SIR 4000 display.

During the investigation, the experiment was divided into two parts, referred to as Experiment 1 and Experiment 2. The main distinction between these experiments was the size of the scanning area. In experiment 1, radargrams were obtained for the initial stage, while in Experiment 2, radargrams were gathered for the subsequent stage. After completing the scans for each phase, the acquired data was transferred to a computer and merged into a three-dimensional model for each phase.

3. Results and discussion

The data was obtained from radargram by using SIR4000 and then was processed by Radan 7 software. 31 radargram was obtained for experiment 1 and 27 radargram was obtained for experiment 2. From the radargram, the average depth was obtained for mild steel pipe. The average depth from the radargram is 1.075 m with a tolerance of ±0.095 m. Compared with the reference depth of 1.15 m, we get the accuracy using GPR is 1.10/1.15 x 100% = 95.65%. For comparison, the 3D file from Radan 7 software shows the pattern of mild steel pipe below the ground (Figure 7-8).

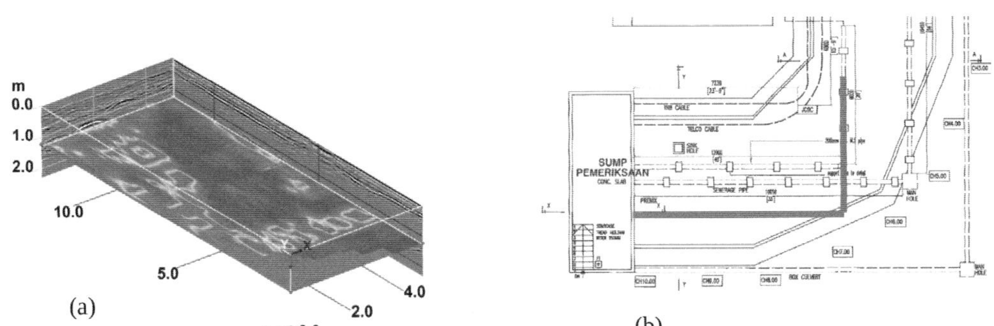

(a) (b)

Figure 7. (a) The 3D view for 1st phase scanning area (b) Contractor's as-built drawing for reference 1st phase.

IOP Conf. Series: Materials Science and Engineering 1308 (2024) 012021 doi:10.1088/1757-899X/1308/1/012021

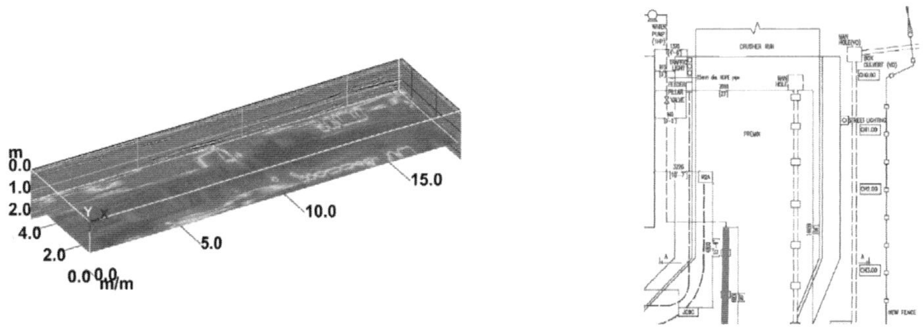

(a) (b)

Figure 8. (a) The 3D view for 2nd phase scanning area (b) Contractor's as-built drawing for reference 2nd phase.

These 3D files were obtained from the combination of radargrams in the X and Y directions. Every radargram has the detail of the image below the surface. From that, using radargram can observe the potential area that has the underground utility that has been desired for inspection. Below are several examples of the radargram:

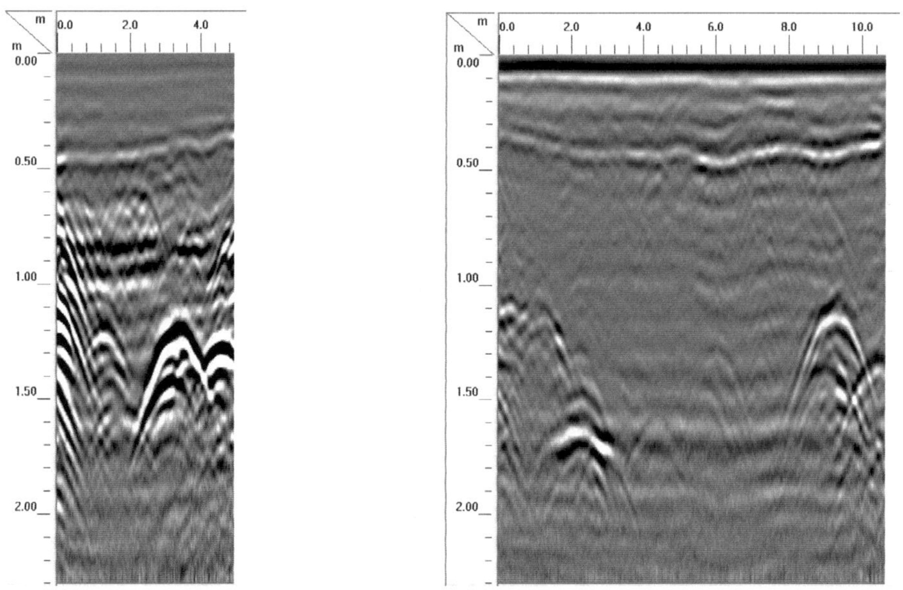

(a) (b)

Figure 9. (a) Radargram for X direction 1st phase (b) Radargram for Y direction 1st phase.

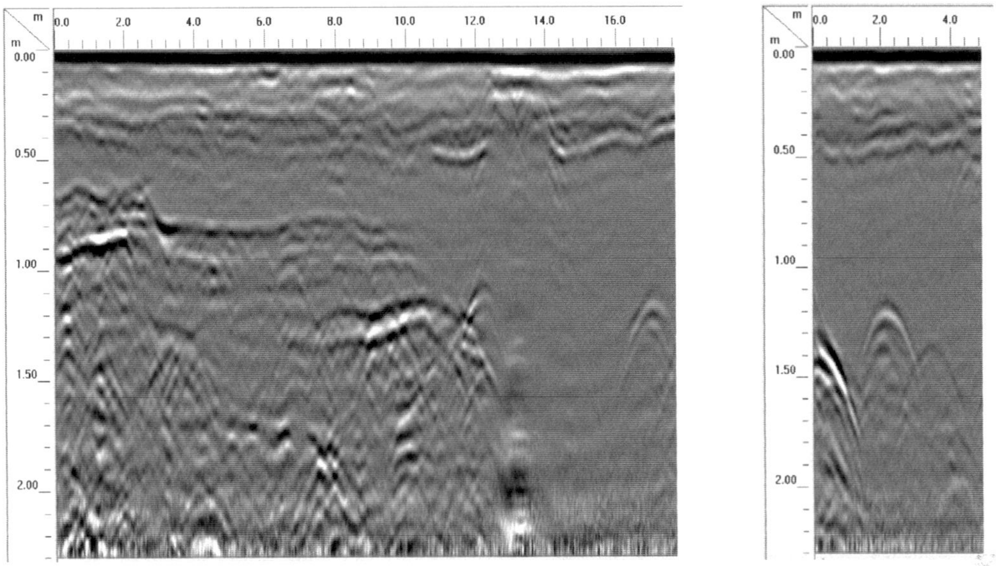

(a) (b)

Figure 10. (a) Radargram for X direction 2nd phase (b) Radargram for Y direction 2nd phase

According to the radargram depicted in Figure 9 and Figure 10, GPR can directly capture images beneath the ground's surface. The radargram presented a depth of approximately 2 meters. Additionally, the distance at which the data is taken makes it easier for the user to identify the targeted area for marking and point reference. As a result, based on data processing, it can be concluded that GPR is a non-destructive, fast, and economical method for evaluating road structures [7]. Through various data acquisition and processing techniques, studies have shown significant success in using GPR to gather information for verifying the state and condition of materials [8,9]. This technology helps overcome the constraint of insufficient cavity samples for 3D radar detection on intelligent learning model training, reduces algorithm training costs, and improves identification accuracy [10,11].

4. Conclusion

The purpose of this study was to identify the location of underground utilities made of mild steel pipes using ground penetrating radar (GPR). By transferring the data obtained from GPR to RADAN 7 software, three-dimensional visualizations can be generated that can be viewed layer by layer for detection. This technology is highly effective in guiding inspectors on where to focus their attention and which areas beneath the ground should be avoided, as it provides a reliable and detailed view of the underground infrastructure. As the technology continue to evolve, it is expected that GPR will become even more sophisticated, leading to further advancements in underground utility inspection.

5. References

[1] Jaw S W and Hashim M 2013 Locational accuracy of underground utility mapping using ground penetrating radar Tunnelling and Underground Space Technol. 35 20–29 doi: 10.1016/j.tust.2012.11.007.

[2] Annan A P 2002 Subsurface sensing technologies and applications *Subsurface Sensing Technol. and Appl.* **3**(4) 253–270 doi: 10.1023/A:1020657129590.

[3] Grote K, Hubbard S and Rubin Y 2003 Field-scale estimation of volumetric water content using ground-penetrating radar ground wave techniques *Water Resour. Res.* **39**(11) doi: 10.1029/2003WR002045.

[4] Chlaib H K, Abdulnaby W and Abd N 2014 Application of the ground penetrating radar to detect weapons caches and unexploded ordnance: laboratory experiments *IOSR J. of Appl. Geol. and Geophys.* **2**(5) 41–50 doi: 10.9790/0990-0254150.

[5] Moffatt D L and Puskar R J 1976 A subsurface electromagnetic pulse radar *Geophys.* **41**(3) 506–518 doi: 10.1190/1.1440629.

[6] Pande J, Singh A, Gupta A, Singh K and Sharma A 2023 Miniaturised ultra-wideband rectangular shaped slot antenna for ground penetrating radar applications *Frequenz* **77**(1–2) 107–114 doi: 10.1515/freq-2021-0318.

[7] Zainal M, Raihan R, Yanis M and Umar M 2023 The mapping of road pavement structure around the seismically active regions by using ground penetrating radar *Key Eng. Mater.* **940** 145–152 doi: 10.4028/p-7mud9g.

[8] Spears M, Hedjazi S and Taheri H 2023 Ground penetrating radar applications and implementations in civil construction *J. of Struct. Integ. and Maint.* **8**(1) 36–49 doi: 10.1080/24705314.2022.2142901.

[9] Li F, Yang F, Qiao X, Xing W, Zhou C, and Xing H 2023 3D ground penetrating radar cavity identification algorithm for urban roads using transfer learning *Meas. Sci. Technol.* **34**(5) 055106 doi: 10.1088/1361-6501/acb6e3.

[10] Webb R W 2017 Using ground penetrating radar to assess the variability of snow water equivalent and melt in a mixed canopy forest, Northern Colorado *Front Earth Sci.* 11(3) 482–495 doi: 10.1007/s11707-017-0645-0.

[11] Azalan M S Z, Esian T, Ali H, Zaidi A F A and Amran T S T 2023 Classification size of underground object from ground penetrating radar image using machine learning technique *J. Phys. Conf. Ser.* 2550 012029 doi:10.1088/1742-6596/2550/1/012029.

Acknowledgments
The authors would like to extent their appreciations and sincere thanks to Ministry of Science, Technology and Innovation (MOSTI), Universiti Malaysia Perlis (UNIMAP) and Agensi Nuklear Malaysia under MOU agreement as well as the Non-Destructive Testing-Material and Structure Integrity Group (NDT-MSI) for the support and co-operation.

AUTHOR INDEX

Institute of Physics
Dirac House, Temple Back
Bristol BS1 6BE UK

ISSN: 1757-8981
ISBN 978-1-7138-9717-0

9 781713 897170

CURRAN ASSOCIATES INC.
proceedings.com